Managing weather-related disasters in Southeast Asian agriculture

This work is published under the responsibility of the Secretary-General of the OECD. The opinions expressed and arguments employed herein do not necessarily reflect the official views of OECD member countries.

This document, as well as any data and any map included herein, are without prejudice to the status of or sovereignty over any territory, to the delimitation of international frontiers and boundaries and to the name of any territory, city or area.

Please cite this publication as:
OECD (2018), *Managing weather-related disasters in Southeast Asian agriculture*, OECD Studies on Water, OECD Publishing, Paris.
http://dx.doi.org/10.1787/9789264123533-en

ISBN 978-92-64-12352-6 (print)
ISBN 978-92-64-12353-3 (PDF)

Series: OECD Studies on Water
ISSN 2224-5073 (print)
ISSN 2224-5081 (online)

The statistical data for Israel are supplied by and under the responsibility of the relevant Israeli authorities. The use of such data by the OECD is without prejudice to the status of the Golan Heights, East Jerusalem and Israeli settlements in the West Bank under the terms of international law.

Photo credits: Cover © Acrylik.

Corrigenda to OECD publications may be found on line at: *www.oecd.org/about/publishing/corrigenda.htm.*

Foreword

In April 2016, OECD Ministers of Agriculture identified that the agriculture and food sector will not only have to respond to the opportunities of increasing demand for food and agriculture products in a sustainable way, but also have to adapt to climate change. To achieve this goal, agriculture and food policies would need to give farmers the means to develop practices that would at once ensure a greater resilience to risk, as well as to cope with more frequent and unpredictable weather-related shocks.

The OECD has made efforts to support such policies by developing recommendations to enable the adaptation of agriculture to climate change and the management of agriculture water risks. In particular, the report *Mitigating Droughts and Floods in Agriculture: Policy Lessons and Approaches* (2016), proposed a comprehensive analysis and key recommendations on policy approaches to the sustainable management of droughts and floods in a sustainable way.

Concurrently, the OECD has taken steps to broaden its expertise in offering robust evidence-based policy advice to emerging partner economies. In the field of food and agriculture, it has developed a partnership with the Association of Southeast Asian Nations (ASEAN) and its member countries to review the policy environments in this region. Born from this collaboration, the report *Building Food Security and Managing Risks in Southeast Asia* (2017), explored effective policy solutions to current and future challenges related to food security in ASEAN countries.

The present work is a follow-up to these two reports. It reviews the policies related to droughts, floods and tropical storms in the agricultural sector for selected ASEAN countries by adapting the general policy framework proposed in Mitigating Droughts and Floods in Agriculture to the local ASEAN context.

The present report was declassified by the OECD Working Group on Agricultural Policies and Markets in December 2017. The report was made possible by a financial contribution from Australia.

Acknowledgements

This report was prepared by the OECD with the active participation of experts and representatives from Myanmar, the Philippines, Viet Nam, and Thailand. Jean-Joseph Cadilhon, Laura Munro and Guillaume Gruère from the OECD Trade and Agriculture Directorate are the main authors. The following OECD experts provided useful suggestions to improve early drafts of the report: Catherine Gamper (Public Governance and Territorial Development Directorate), Rachel Scott (Development Co-operation Directorate), Leigh Wolfrom (Directorate for Financial and Enterprise Affairs), Emily Gray, Jared Greenville, Shingo Kimura, Andrzej Kwiecinski, and Olga Melyukhina (Trade and Agriculture Directorate). Malory Greene and Tiyarat Niamkohphet-Cader (Global Relations Secretariat) helped identify government contacts in the countries studied. Administrative and editing services were provided by Nathalie Elisseou Leglise, Stéphanie Lincourt and Theresa Poincet from the Trade and Agriculture Directorate. The authors would also like to thank Natasha Cline-Thomas, Robert Akam, and Michèle Patterson for substantive and editorial comments on the report.

In addition to the experts and representatives interviewed during the field visit, special thanks are extended to the collaborators who provided assistance in organising the field visit and who co-ordinated the review of earlier drafts of this report in the countries studied: Tisha Pia E. De la Rosa (Department of Agriculture, Philippines), Dinh Pham Hien (Ministry of Agriculture and Rural Development, Viet Nam), Thanda Kyi (Ministry of Agriculture, Livestock and Irrigation, Myanmar), Pham Quang Minh (Food, Agriculture and Forestry Division, ASEAN Secretariat) and Siriporn Thanaratchataphoom (Ministry of Agriculture and Cooperatives, Thailand).

Table of contents

Tables

Figures

Boxes

Follow OECD Publications on:

Acronyms

ADB	Asian Development Bank
AADMER	ASEAN Agreement on Disaster Management and Emergency Response
ASEAN	Association of Southeast Asian Nations
APTERR	ASEAN Plus Three Emergency Rice Reserve
CRED	Centre for Research on the Epidemiology of Disasters
DDPM	Thai Department of Disaster Prevention and Mitigation
DMH	Myanmar Department of Meteorology and Hydrology
DRM	Disaster risk management
DRR	Disaster risk reduction
EM-DAT	Centre for Research on the Epidemiology of Disasters' emergency events data base
FAO	Food and Agriculture Organization of the United Nations
FCRI	Viet Nam Field Crops Research Institute
ha	Hectare
GDP	Gross domestic product
IPCC	International Panel on Climate Change
MoAC	Thai Ministry of Agriculture and Co-operatives
MOALI	Myanmar Ministry of Agriculture, Livestock and Irrigation
MADB	Myanmar Agricultural Development Bank
MARD	Viet Nam Ministry of Agriculture and Rural Development
MAPDRR	Myanmar Action Plan on Disaster Risk Reduction
NDRRMC	Philippines National Disaster Risk Reduction and Management Council
NLUA	Philippines National Land Use Act
PAGASA	Philippine Atmospheric, Geophysical and Astronomical Services Administration
PDNA	Post-Disaster Needs Assessment
RIMES	Regional Integrated Multi-Hazard Early Warning System for Africa and Asia
UN	United Nations

Executive summary

The exposure and vulnerability of the agricultural sector to weather-related disasters such as storms, droughts, or floods is increasing. Climate change is likely to exacerbate the extreme nature and frequency of such disasters, with potentially catastrophic impacts on the agricultural sector in several countries of the Association of Southeast Asian Nations (ASEAN). Four ASEAN countries are ranked among the countries most affected by weather-related disasters worldwide due to their geographical location within the tropics: Myanmar, the Philippines, Thailand, and Viet Nam. Their governments have tended to focus on response and recovery from successive disasters, but the recent milestone weather-related disasters that have inflicted massive damage and losses to their agricultural sector, points to the need for greater prevention, mitigation and preparation measures.

As smallholder farmers account for a substantial part of the rural population in these four countries – as well as important contributors to national food security – an appropriate policy environment to tackle these weather-related risks is critical. Furthermore, all four countries are currently, or have the potential to become, major players in international agricultural commodity markets, in particular for rice. Providing relevant policy advice to ASEAN countries on limiting the agricultural sector's exposure to these risks is important for these countries, but also globally as it can help limit the volatility of international commodity markets.

This report aims to identify good policy practices and to provide evidence-based policy advice to strengthen the ASEAN agricultural sectors' resilience to weather-related disasters. It reviews policy approaches to droughts, floods and tropical storms in Myanmar, the Philippines, Thailand and Viet Nam, which are then compared with good practices as highlighted by OECD recommendations on mitigating risks from droughts and floods in agriculture, governance for strategic disaster risk reduction and management, and disaster-risk financing strategies and tools. In particular, the report assesses policy instruments to cope with catastrophic risk (classified as risks where individuals and markets are not equipped to respond to the large scale of the disaster) across the risk management spectrum: (i) prevention and mitigation; (ii) preparedness; (iii) response; and (iv) recovery.

The four countries studied have the legal and regulatory frameworks for disaster risk management based on international guidelines and good practices. However, the major challenge they face is to co-ordinate the activities of the different relevant governmental institutions at national and local levels to implement these frameworks. As a result, the provision of emergency supplies to affected areas, such as food and drinking water, along with agricultural inputs and equipment needed to recover from a disaster, can be slow.

Nevertheless, agriculture is usually well integrated in government disaster risk management activities given its relative importance in these countries' economies. Ministries in charge of agriculture are involved in many dimensions of disaster risk management planning and hold a key role in disaster response activities in rural areas or if any water management infrastructure are involved: dams, levies, or irrigation and drainage networks.

Large water infrastructure projects are a priority for government spending in the region to reach agricultural – in particular, rice – production objectives. This is not coherent with mitigating the impact of extreme-weather events: irrigated farmers, for example, might be encouraged to grow water-thirsty crops, like rice, in areas not naturally suited for such crops, but where it has been made possible by irrigation development.

Awareness among Southeast Asian farmers of practices to reduce their risk exposure is relatively low, and technical advice on how to strengthen resilience is variable. This is due in part to the variable quality of extension advice, which depends on the capacity of the local extension agent, the level of remoteness of the farming community, and the civil society and agribusiness networks that can help transfer knowledge and information. Many on-farm practices and tools that increase resilience could be disseminated without further need for research.

Some economic incentives in ASEAN countries encourage farmer resilience, but many measures in the countries reviewed also counter these policies and increase farmers' vulnerability to weather-related disasters. Of particular concern are agricultural support measures that distort farmer incentives. Government-provided agricultural insurance also exists in some ASEAN countries, though current policies limit resilience due to design flaws and a focus on water-thirsty crops.

The information systems for weather and water levels are relatively well developed at the watershed level in the countries reviewed. These systems could still be strengthened to improve the geographical coverage of data collection and to provide timely and relevant information allowing farmers to prepare against oncoming weather-related disasters.

The use of financial tools to respond and recover from weather-related disasters is varied. Most low-income households in the region do not have precautionary savings they can use if they are hit by a weather-related disaster. Many farmers in the region are also in a continuous state of debt. Thus, several countries have set up disaster-linked cash transfers providing some compensation in cash to farmers affected by disaster, but the effectiveness of such schemes could be improved. Furthermore, all four countries also resort frequently to debt rescheduling, a useful financial tool to aid the recovery process though it may increase farmer vulnerability: in some countries, interest can be written off for certain loan programmes. Finally, given farmers' limited savings, all four countries provide in-kind agricultural inputs and equipment in areas affected by a disaster.

Key recommendations

- Strengthen the prevention and mitigation components of disaster risk management by aligning policy incentives and by integrating environmental resilience into infrastructure planning and extension systems.
- Implement and enforce water allocation and water use restriction instruments like water pricing to steer farmers towards more efficient use of water resources.
- Improve the co-ordination of government and partner institutions' activities to enable a more timely response to disasters.
- Improve the timely distribution of inputs, equipment and social protection measures like disaster-linked cash transfers to strengthen farmers' capacity to recover from disasters.

Chapter 1. Overview and policy recommendations

This chapter summarises the findings and recommendations of Managing Weather-Related Disasters in Southeast Asian Agriculture. It emphasises the need for the reviewed countries to improve their disaster risk management institutions, and to strengthen their prevention and mitigation policies to reduce agriculture's risks to weather-related disasters. It then outlines recommendations to improve preparedness, crisis management and recovery measures in these four countries and identifies three priority actions in each country.

The exposure of Southeast Asia to increasingly frequent and intense weather-related disasters is a growing concern for agricultural producers. According to the latest reports of the Global Climate Risk Index (2017), four member countries of the Association of Southeast Asian Nations (ASEAN) currently rank among the most affected countries worldwide in terms of the impact of past extreme weather events: Myanmar (2nd worldwide), the Philippines (4th in 2016 and 5th in 2017), Viet Nam (7th in 2016 and 8th in 2017), and Thailand (9th in 2016 and 10th in 2017). In the past 20 years, the Philippines, Viet Nam and Thailand have all suffered more than one catastrophic flood, while Myanmar had a 95% probability of being hit by a disastrous flood. These weather-related disasters have had catastrophic impacts on agriculture, posing a threat to the region's food security and further development potential.

As many ASEAN countries strive to meet national objectives of increasing agricultural productivity and production in the coming years, strengthening resilience will be key. While trade-offs may exist in the short term, a careful assessment of policy implications for productivity *and* resilience will reveal that investment in resilience will generate productivity payoffs in the long-term.

This report reviews policy approaches to droughts, floods and tropical storms in selected Association of Southeast Asian Nations (ASEAN) countries (Myanmar, the Philippines, Thailand and Viet Nam) in an effort to identify good practices and strengthen the resilience of the agricultural sector across the region. Chapter 2 builds the case for improved disaster risk management against droughts, floods and tropical storms in Southeast Asia. Chapter 3 then provides evidence of the high risk exposure of the agricultural sector in the region to these weather-related disasters. Finally, Chapter 4 reviews the policies in place in the selected ASEAN countries to manage risks to agriculture from weather-related disasters. In particular, policy instruments across the risk management spectrum – (i) prevention and mitigation, (ii) preparedness, (iii) response and (iv) recovery – are discussed and assessed using the benchmark of OECD recommendations and guidelines, in particular the OECD framework developed from policy lessons and approaches to mitigate droughts and floods in agriculture. Based on these elements, the report draws a number of conclusions and policy recommendations for ASEAN countries and the ASEAN Secretariat, which are summarised in this chapter.

1.1. Improving disaster risk management institutions to limit damages and losses from weather-related disasters in ASEAN countries' agricultural sector

ASEAN countries have set up regulatory frameworks and institutional settings for disaster risk management (DRM) according to international recommendations and good practices. The ministries and institutions in charge of agriculture are actively involved in this set up. However, to date the policy emphasis has been mainly placed on coping with, and recovering from disasters, rather than preventing, mitigating and preparing for future disasters. Improvements to inter-ministerial co-ordination could also be made. DRM is largely implemented top-down: instructions are developed at the national level and carried-out at the local level. ASEAN countries should consider providing local governments with resources and capacity to enable them to implement DRM measures that are pertinent to their specific situation; this would support integration of different policy instruments in locally appropriate policy packages and facilitate collaboration between different local administrations involved in DRM.

- *Strengthen disaster risk reduction planning.* Co-ordinated strategic thinking between all government departments and partner institutions involved in DRM should help reduce the sector's risk exposure, thus reducing the cost of response and recovery measures.

- *Improve the communication channels between government departments involved in DRM.* This will allow better co-ordination in planning to translate into better co-ordination when responding to disasters.

1.2. Enhancing prevention and mitigation policy instruments to reach OECD good practices

While ASEAN risk management efforts have focused primarily on response and recovery efforts, the four countries studied have taken preliminary steps to reduce their risk exposure. According to OECD policy lessons on mitigating the risk of droughts and floods to agriculture, key prevention and mitigation measures include: (i) strengthening the resilience of water-related infrastructure projects; (ii) aligning regulations; (iii) increasing farmer awareness; (iv) aligning economic incentives to reduce vulnerability; and (v) designing insurance policies to encourage resilience.

Large water infrastructure projects are a priority for government spending in the region to reach agricultural production objectives. The water infrastructure in many ASEAN countries has contributed to expand the land coverage of rice production, but is not as well suited to deal with a changing climate, and more frequent and intense weather-related disasters. Infrastructure projects are centrally planned with little consultation of local stakeholders. Their planning is only starting to account for their associated environmental costs and benefits in line with the higher risk of weather-related disasters and other objectives of longer-term sustainability. What is more, their production objective is not coherent with mitigating the impacts of extreme-weather events on the agricultural sector as irrigated farmers might be encouraged to grow water-thirsty crops like rice in areas that are not naturally suited for these crops, but where it has been made possible by irrigation development.

- *Undertake wide stakeholder consultations and include environmental cost-benefit and risk assessments for water infrastructure projects.* Planning for new water infrastructure or refurbishing existing ones should follow standard international recommendations, including an environmental risk assessment to address the increased risk of climate change and weather-related disasters. It should ideally involve the local stakeholders and users to adjust the infrastructure to the needs of the local territory and its communities. To improve the sustainability of these projects, investments should be coupled with a higher charge to users benefiting from the infrastructure to recoup some of the investment costs and at least reflects the actual costs of maintenance.

- Reconsider the objective of expanding actual irrigated land area and *concentrate on improving the water efficiency of current irrigation infrastructure* to save water resources. For land that is not yet irrigated, and based on the findings of an environmental cost-benefit analysis of new infrastructure, governments should consider the less costly policy option of extending technical innovations to farmers that allow them to produce non-traditional commodities to ensure their income and food security.

Awareness among Southeast Asian farmers of practices to reduce risk exposure is low, and the technical advice offered to them on how to strengthen resilience is variable. While extension services are encouraging crop diversification in many countries, other on-farm practices to strengthen resilience – such as the adoption of resistant seed varieties and effective water management – are less widely promoted. This is due in part to the variable quality of extension advice, which depends on the capacity of the local extension agent, the level of remoteness of the farming community, and on the civil society and agribusiness networks that can help transfer knowledge and information. In some cases, it is also driven by a limited resilience focus for research and development, although many on-farm practices and tools that increase resilience could be disseminated without further research.

- *Mainstream disaster risk management in ASEAN countries' agricultural extension curricula.* Existing technical assistance programmes should provide information about farmer's risk exposure and affordable on-farm strategies to reduce it. Such strategies could include crop diversification, but should also extend to farm practices such as harvesting rainwater in drought-prone regions and natural flood barriers in flood-prone regions. In the more remote areas where extension officers are sparse and communication networks patchy, collaboration with international development partners and local innovation systems would help disseminate information about risk exposure and agricultural innovations to isolated farmers.

- *Measure farmer awareness about risk exposure and risk management.* Statistical departments in ASEAN countries should collaborate to add one or two questions in regular agricultural surveys to gather data that can be compared across countries related to farmers' understanding of how their practices affect their resilience. This information would help improve the monitoring of the impact of extension services on farmers' awareness of climate-risky practices.[1]

ASEAN countries can strengthen policies and regulations to improve collective and individual use of water in the agricultural sector. At present, there are only a few mechanisms to allocate water rights and regulate water use. To regulate water volume use, the ASEAN countries studied rely mainly on less effective land zoning measures and the promotion of specific crops adapted to the current climatic situation in a given agro-ecological setting by farmers, rather than limiting their water use with more direct policy instruments.

- *Adjust the regulatory framework to steer towards more efficient use of water resources by the agricultural sector.* A comprehensive and coherent set of policies could include some of the following: allocate water use rights to individual or collective users; develop water use restriction measures to be implemented in times of crisis; allow water rights to be auctioned and transferred.

- *Invest in water meters.* Better water meters would gather better evidence to elaborate policy instruments that regulate water use more directly than would zoning and farm practice suggestions. Given their limited financial resources, ASEAN countries should collaborate with the private sector and the international development community to co-finance investments in water use measurement tools.

Although several economic incentives that encourage resilience exist (e.g. preferential credit for efficient irrigation schemes), many agricultural support measures in Southeast Asia counter these policies and increase farmers' vulnerability to weather-related disasters. Market price support measures (such as import restrictions and minimum prices), preferential insurance subsidies, and concessional loan terms for rice farmers are of particular concern.

- *Reduce economic incentives that increase vulnerability.* Market price support measures and preferential subsidies should be removed to avoid encouraging farmers towards agricultural products that are not resilient in local conditions. Water pricing schemes should be revised to reflect the full cost of water delivery and should be accompanied by resource-decoupled compensation, where appropriate and applicable. Such reforms would help reduce overuse of water in drought-prone regions. If governments must take the decision to restrict water extraction in times of extreme drought or to flood agricultural land to protect the economic interests and assets of other sectors of the economy, farmers should be adequately compensated for their losses.

Government-sponsored agricultural insurance exists in some Southeast Asian countries in the absence of private sector provision. It usually covers various weather-related disasters for the main commodities grown, but current policies limit resilience due to design flaws and a focus on water-thirsty crops. The Philippines and Thailand are more advanced and offer indemnity-based policies for selected agricultural products. Although both schemes offer advantages, they also have design constraints that limit the potential benefits in terms of risk reduction and risk transfer. Myanmar and Viet Nam have yet to develop agricultural insurance products at the national scale, although several pilot projects have been initiated.

- *Develop cost-effective insurance programmes for catastrophic risks that provide pay-outs in a timely manner.* Hybrid insurance products, for example, are appropriate in many contexts as they can include: (i) an index-linked partial pay-out (triggered by shocks recorded at an aggregate level) to help farmers face immediate costs after a disaster; and (ii) an indemnity component based on losses recorded at the farm level. The index component helps to reduce administrative costs for the government and provides quick support to farmers, while the indemnity component ensures that the pay-out more closely reflects the losses suffered by farmers.

- *Price the programme so that it ensures adequate take-up levels, but creates minimal distortions to farmer incentives to invest in resilient practices.* The experience of government-provided crop insurance is that subsidies are needed to increase take-up among farmers with financial constraints. However, these subsidies should be structured in a way that does not increase vulnerability by distorting farmer incentives to reduce risk. For instance, subsidies should not be limited to specific crops, as this may discourage farmers from planting more resilient alternatives.

- *Couple insurance with educational programmes about how the policy works.* Capacity development should allow farmers to better understand which risks are covered and which are not, helping them adopt strategies to reduce their risk exposure. Such information campaigns would also help to encourage take-up.

1.3. Efforts to strengthen institutional and household preparedness are starting to bear fruit

Many ASEAN countries have taken an important first step in preparedness by developing efficient weather and hydrological information systems. As is the case for extension services, getting relevant and timely information on incoming weather-related disasters to individual farmers in all areas that could be affected continues to be a challenge. To improve the preparedness of farm households to incoming weather-related disasters, the ASEAN countries under study are improving the coverage of their weather and water information systems. However, a gap remains in terms of conveying relevant and timely hydrometeorological information to the farmers.

- *Improve communication between the system and farmers.* ASEAN countries should improve the information and communication tools and processes to ensure hydrometeorological information is effectively conveyed to the end users: individual farmers. A particular effort will be needed to extend coverage of the information system in remote rural areas.

- *Continue investing in hydrometeorological information systems.* ASEAN countries should continue their combined technical collaboration, as well as collaborate with the international development community, to improve and invest in measurement equipment, data management systems, and local innovation networks. This would provide reliable information to policy makers as they develop policy, to insurance providers and markets that assess farmer risk, as well as to farmers and rural communities thus allowing them to strengthen their resilience to weather-related shocks. Given the limited financial resources of ASEAN countries, their governments should seek co-financing from the private sector and concessional loans from the international development community.

A preparedness measure at the household level – precautionary savings –helps wealthier farmers to manage risk in Southeast Asia. Given low and volatile income levels in many ASEAN countries, the majority of smallholder farmers do not have access to precautionary savings. However, among farmers with higher incomes, savings are commonly set aside to prepare for emergencies such as weather-related disasters. Most farmers keep their savings at home, but the use of formal savings accounts is increasing.

- *Encourage precautionary savings in formal settings.* The use of formal savings mechanisms should continue to be encouraged across the ASEAN region; for example, by coupling access to credit and other financial services to the opening of a savings account. It is important to note that subsidies initially needed to encourage farmer take-up of insurance products might incentivise lower precautionary savings, pointing to the importance of co-ordinating the various financial incentives and support products available to farmers to help them manage their risk.

1.4. Adapting to and learning from high frequency weather-related disasters to manage crises in the ASEAN region

Many ASEAN countries have aligned their crisis management procedures with international best practice guidelines. Thanks to the regulatory frameworks and institutions in place, ASEAN countries can respond to a crisis, and provide emergency relief for farmers and rural areas to cope with a disaster. However, the lack of co-ordination between different governmental services at the national and local levels can create hurdles to prompt government response to disasters. Additional evidence gathered from some of the countries studied points to the need for better functioning government procurement systems to buy the necessary equipment and first-necessity products in large quantities and to cope with inflows of funds from international development partners, with adjoining procurement rules and regulations.

- *Improve the co-ordination mechanisms between governmental departments and local authorities to allow more fluid response to crisis at local level.* The devolution efforts concerning powers, budgets and human resources witnessed in the Philippines, Thailand and Viet Nam are probably an appropriate solution to improve local response capacity. However, better governance and co-ordination between local units of the line ministries is needed to allow this devolution process to bear fruit for rapid disaster response.

- *Streamline governmental procurement systems.* Governments in the ASEAN countries under study are the first point of call to respond to a disaster. They are therefore expected to have the appropriate systems allowing them to procure and distribute emergency goods and equipment. More efficient government procurement systems would allow them to respond more rapidly to disasters and make a better use of the technical and financial support often proposed by the international development community when disasters hit.

Disaster-linked cash transfer programmes, a second tool supporting the response phase, are already employed in some Southeast Asian countries but many of these programmes are slow to make disbursements and costly for the government. In some countries, loss assessments are undertaken for each farm, which slows the disbursement process and inflates programme costs. In certain countries, all farmers are eligible regardless of farm size and income levels; this is problematic as it may actually increase risk-taking among wealthier farmers that have other forms of insurance. It also increases costs for governments, which are likely to persist in the long term: offering disaster-linked transfers to all farms fosters an attitude of dependence and reducing recourse by larger farms to insurance due to their expectation of support.

- *Design disaster-linked cash transfers to offer quick, targeted and cost-effective support to low-income farmers.* As take-up of agricultural insurance products tends to be very limited among low-income households, disaster-linked cash transfer programmes should be designed to fill this gap in financial coverage. By collecting satellite data of agricultural losses, such programmes can disburse quick payments via mobile banking and ensure that beneficiaries (farmers whose income is too low to afford the purchase of insurance covering weather disaster risk) receive a cash transfer within a few weeks after a disaster. In order to reduce the financial burden on the government and societal expectation that the government will step in for all farmers, such a programme should be limited to support low-income farmers who cannot afford insurance products.

1.5. Extending recovery measures beyond pre-disaster conditions and building resilience against future disasters

In terms of post-disaster recovery by the agricultural sector, the countries studied have the mechanisms to allow agricultural inputs and equipment to be distributed following a disaster. The majority of countries studied have set up a national reserve, sometimes relayed by regional or provincial reserves, of relevant equipment and farm inputs that are needed to help farmers recover quickly from a disaster. However, some administrations could improve their procurement procedures to allow timely spending of the funds allocated to respond to the disaster.

- *Make complementary use of a national reserve, local and international markets to source inputs and equipment for recovery after a disaster.* Such reserves are useful mechanisms for rapid deployment of essential farm inputs to prepare a new planting season or to sustain livestock numbers. However, given the improving transport infrastructure in many ASEAN countries, and the increasing market integration between ASEAN countries and their regional partners, it is worth exploring the possibilities that local, regional and international markets might bring to source inputs and equipment more efficiently than would a national reserve mechanism.

The four countries studied frequently resort to debt rescheduling, a common financial tool to aid the recovery process. In some countries, loan interest can be written off for certain programmes, which may encourage high-risk investments and thus increase vulnerability. However, formal savings accounts are still not widely used, which limits the scope of this policy measure to wealthier farmers with formal loans.

- *Design debt rescheduling measures to encourage resilience through clear conditions and limited scope.* Agricultural loans should include clear conditions under which debt is rescheduled. This provides farmers with a limited, implicit "insurance" against catastrophic risks and enables them to make productive investments. At the same time, such rescheduling terms should be conservative and avoid write-offs of debt, whether it concerns the interest or principal. Governments should align current public loan programmes with such standards and introduce regulation to ensure private loans also comply.

More fundamentally, recovery measures are essentially geared to helping the agricultural sector get back to the situation where it was before the disaster struck. Recovery measures are not fostering the institutional and structural changes needed to make agriculture more resilient to future disasters.

- *Take a longer-term view towards improving agriculture's resilience to frequent disasters.* With the help of international organisations specialised in policy analysis and reform, ASEAN countries and the ASEAN Secretariat should elaborate policy instruments to help steer the agriculture sector to better prepare for future weather-related disasters. In the event of a disaster, once immediate emergency needs have been attended to, governments should review the lessons learned from their past response and review the checklist of policy approaches for disaster prevention, mitigation and preparedness to help the sector be better prepared for a future weather-related disaster. This closes the loop of the DRM cycle, as recommended by international good practices.

1.6. Country-specific priority actions

This report has identified a current state of policy measures on DRM in general, and on tackling the risk to agriculture from droughts, floods and tropical storms in particular, in Myanmar, the Philippines, Thailand and Viet Nam. Given the evidence gathered, the governments of these countries could consider taking the following priority actions to bring their agricultural risk management systems in line with OECD and international good practices. Several recommendations are similar because the countries face similar gaps in their policy environment.

Myanmar

- Mainstream disaster risk management into existing agricultural extension initiatives by exploring innovative, low-cost tools that reach farmers with risk management advice.

- Given that most of the country's water dams, dykes and irrigation perimeters were established many years ago, review the calibration of all major water infrastructures to take account of the increased frequency and larger potential impact of future weather-related disasters.

- Explore opportunities to build on existing cash transfer mechanisms by including a disaster-linked component that issues timely payments after extreme events. Collaborate with experts in the international development community to ensure that such a programme is carefully targeted and designed to strengthen resilience of low-income farmers.

Philippines

- Mainstream disaster risk management into existing agricultural extension initiatives by exploring innovative, low-cost tools to reach farmers with risk management advice.

- Given the localised nature of floods and typhoons affecting the country, agree on specific local DRM objectives with Local Government Units to encourage the effective devolution of DRM to the administrative level that will be best placed to prevent and respond to these risks.

- Given the distortive effect of current market price support measures on rice production, review existing policies currently favouring rice to allow more resilient crops to be chosen by farmers according to their local market and agro-ecological conditions.

Thailand

- Mainstream disaster risk management into existing agricultural extension initiatives by exploring innovative, low-cost tools to reach farmers with risk management advice.

- Promote efficient water use through carefully-designed water pricing measures: charge farmers' use of irrigation and groundwater at a rate that reflects the costs of investment in infrastructure and delivery of this essential natural resource to encourage efficient usage.

- Given the distortive effect of current market price support measures on rice production, review existing policies currently favouring rice to allow more resilient crops to be chosen by farmers according to their local market and agro-ecological conditions.

Viet Nam

- Mainstream disaster risk management into existing agricultural extension initiatives by exploring innovative, low-cost tools to reach farmers with risk management advice.

- Promote efficient water use through carefully-designed water pricing measures: charge farmers' use of irrigation and groundwater at a rate that reflects the costs of investment in infrastructure and delivery of this essential natural resource to encourage efficient usage.

- Given the distortive effects of commodity-specific policies, refrain from applying policies favouring rice to allow more resilient crops to be chosen by farmers according to their local market and agro-ecological conditions.

Note

[1] In the World Bank's Living Standards Measurement Study survey for Viet Nam, the sections 9.4.5 on the reasons for changing cropping structures and 9.4.6 on access to agricultural promotion services provide examples of relevant questions that can help measure progress in farmer awareness to weather-related risk (LSMS, 2017).

Reference

LSMS (2017) "Viet Nam household living standards survey 2004", Living Standards Measurement Survey website, http://iresearch.worldbank.org/lsms/lsmssurveyFinder.htm (accessed 24 May 2017).

Chapter 2. Droughts, floods, tropical storms and Southeast Asian agriculture: Building the case for improved disaster risk management

This chapter provides evidence on the high exposure and vulnerability to weather-related disasters of the agricultural sector of Southeast Asian countries. It then presents the relative importance of the agricultural sector in the economy of these countries. Finally, the chapter summarises the policy analysis framework used to review policy measures managing drought, flood and tropical storm risk in agriculture.

2.1. The threat posed by weather-related disasters to Southeast Asian agriculture

The agricultural sector's exposure and vulnerability to weather-related disasters is significant and increasing. The Food and Agriculture Organization of the United Nations (FAO) has estimated that 25% of all damage caused by such disasters in 48 developing countries between 2003 and 2013 affected the agricultural sector (FAO, 2015a). In particular, droughts, wild fires, cyclones, floods and cold waves led to agricultural land becoming temporarily unproductive, loss of livestock, a greater prevalence of pests, and reduced crop production. According to this same review, 84% of the economic impact of droughts was borne by agriculture during this same time period (FAO, 2015a). Another study estimated that severe droughts and extreme heat between 2000 and 2007 were responsible for crop losses equivalent to 6% of global cereal production (Lesk et al., 2016). Such losses are projected to increase as events become more frequent and severe, but also less predictable, due to climate change (Field et al., 2014). It is also likely that the impact of weather-related disasters will be concentrated in a limited number of geographical hotspots with potentially strong impacts on agriculture and world food security (OECD, 2017c).

Drought, flood and tropical storm risks[1] are of particular concern for the agricultural sectors of several countries in the Association of Southeast Asian Nations (ASEAN) (FAO, 2015a). According to the latest reports of the Global Climate Risk Index (2017) ranking countries based on the impact of past extreme-weather events like floods and heat waves in the past 20 years, Southeast Asia has been one of the most exposed regions to weather-related disasters. Four ASEAN countries rank among the most affected worldwide and will be given particular attention in this study: Myanmar (2nd worldwide), the Philippines (4th in 2016 and 5th in 2017), Viet Nam (7th in 2016 and 8th in 2017) and Thailand (9th in 2016 and 10th in 2017). The Global Water Partnership and OECD highlight similar trends. For instance, Myanmar, Thailand and Viet Nam are among the top ten countries for flood risk exposure (Sadoff et al., 2015). Moreover, meteorological data from Myanmar show strong evidence of the country's growing vulnerability to climate change with rising temperatures, shorter monsoon seasons, and more frequent episodes of intense rainfall and severe cyclones along the coastline (ADB, 2013; Raitzer et al., 2015). Given the key role of the agricultural sector in Southeast Asia, both for food security as well as for development (Box 1), a more comprehensive response to these weather-related disasters is needed. In ASEAN, recent major catastrophic events like the 2008 Nargis cyclone in Myanmar, the 2011 floods in the Central valley of Thailand, the 2013 typhoon Haiyan in the Philippines and the 2015-16 *El Niño* droughts in southern Viet Nam, have highlighted the impact of weather-related disasters on the agricultural sector. Table 3.1 provides evidence of the damages and losses to agriculture caused by these milestone disasters.

The agricultural sectors of Myanmar, the Philippines, Thailand and Viet Nam not only face a growing external threat from climate change, their resilience to catastrophic weather events has also been diminished by internal developments. For example, deforestation in the past century has made the region much more susceptible to floods and droughts; upland deforestation in Myanmar and Viet Nam continues to degrade watersheds (FAO, 2016). Should the People's Republic of China (hereafter "China") and the Lao People's Democratic Republic located upstream finalise all their plans for dams, the water supply to downstream Cambodia, Myanmar, Thailand and Viet Nam might be jeopardised. The Mekong River Commission has made an important contribution to tackling this challenging issue, but there is still work to be done to ensure sustainable management of the Basin.

Building on the OECD's past work on risk management of droughts and floods in the agricultural sector (OECD, 2016b), this report analyses drought, flood and tropical storm risk management policies and producer strategies in a selected number of ASEAN countries with high risk exposure: Myanmar, the Philippines, Thailand and Viet Nam. As such, this report provides useful information and lessons for other parts of the world facing similar risks.

Box 2.1. Agriculture remains an important economic sector in Southeast Asian countries

As a net exporter of agricultural commodities and food products, the Association of Southeast Asian Nations (ASEAN) has become an important global player in agricultural production, thereby contributing to global food security. Myanmar, Thailand and Viet Nam are the top three regional rice exporters. Other key agricultural exports for Myanmar, the Philippines, Thailand and Viet Nam are fresh horticultural produce, nuts and meat. Perennial crops, such as coffee and rubber, are also important agricultural commodities in Thailand and Viet Nam.

The structure of the agricultural sector in ASEAN countries is dominated by small land holdings managed by households. Accordingly, the number of agricultural holdings in ASEAN countries is higher than that in comparable OECD countries. Likewise, the average farm size is much smaller: between 2 and 3.2 ha per farm in Myanmar, the Philippines and Thailand, and only 0.7 ha per farm in Viet Nam.

The agricultural sector still accounts for a large share of employment in some ASEAN countries, though its contribution to gross domestic product (GDP) is typically much lower. In 2014 agriculture employed 30-47% of the labour force in the Philippines, Thailand and Viet Nam. At the same time, contributions to GDP ranged from 10% to 18% in the same countries. Agriculture holds a bigger share of Myanmar's economy: 70% of the labour force and 37.8% of GDP (FAO-MM, 2017). Agriculture is thus an important priority sector for development and poverty reduction measures in the region.

Figure 2.1. Number of agricultural holdings and average size

Estimates during the 2000.

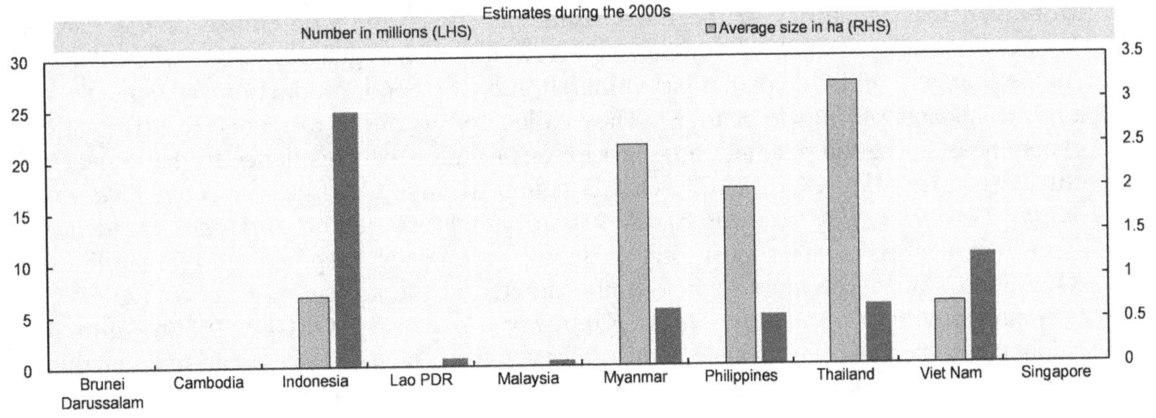

Note: Estimates for each country relate to data collected during the 2000s. Specifically, Indonesia (2003), Malaysia (2005), Myanmar (2010), the Philippines (2002), Thailand (2003) and Viet Nam (2001)..
Source: Lowder et al. (2014)..

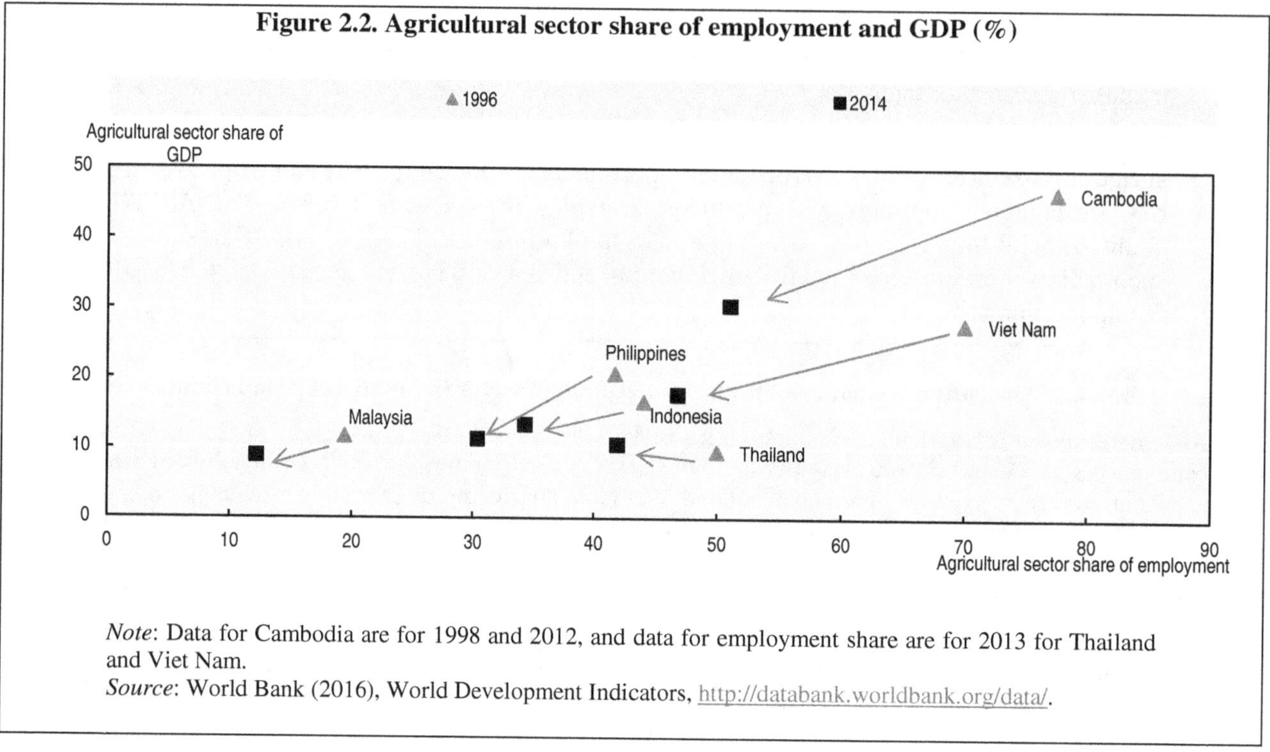

Note: Data for Cambodia are for 1998 and 2012, and data for employment share are for 2013 for Thailand and Viet Nam.
Source: World Bank (2016), World Development Indicators, http://databank.worldbank.org/data/.

2.2. A policy analysis framework for mitigating weather-related risks

A growing recognition and momentum has emerged in the international community about the importance of managing the risk of weather-related disasters. Building upon the good practices in disaster risk management of various countries, the Sendai Framework for Disaster Risk Reduction 2015-30, for instance, aims to steer countries towards increased resilience to risks linked to disasters, including weather-related ones like droughts, floods and storms (UN, 2015). The new sustainable development agenda is similarly aligned; sustainable development goal 13 aspires "to strengthen resilience and adaptive capacity to climate-related hazards and natural disasters in all countries" (Alano and Lee, 2016). Moreover, the signatories of the Paris Agreement from the Conference of the Parties' 21st Session recognised "the fundamental priority of safeguarding food security and ending hunger, and the particular vulnerabilities of food production systems to the adverse impacts of climate change". They called for enhanced exchange of information, experiences, and good practices amongst Parties to raise their resilience to the impacts of climate change (UNFCCC, 2015). OECD countries have also issued *Recommendations on the Governance of Critical Risks*, which set global policy guidelines to identify, assess, reduce risks and boost preparedness, emergency response and rehabilitation (OECD, 2014). Furthermore, although not targeted to the agricultural sector, the OECD *Recommendation on Disaster Risk Financing Strategies* (OECD, 2017b) provides guidance on the financial management of disaster risks, including on creating an enabling environment for risk transfer and the management of public exposures to disaster risks.

Managing such risks in the agricultural sector requires a comprehensive and targeted set of anticipatory (*ex ante*) and post-crisis (*ex post*) policies (OECD, 2016b; OECD, 2009; UNISDR and CRED, 2015). After having identified the hazards and risks related to weather-related disasters facing the agricultural sector, *ex ante* policy tools include: (i) prevention and mitigation measures to reduce risk exposure and (ii) preparedness measures to cope with risks that cannot be eliminated. *Ex post* tools include: (iii) response measures to minimise effects in the short term and (iv) recovery measures to facilitate "bouncing back" in the long-term. As illustrated in Box 2.2, a range of measures are available across these areas and require varied degrees of government involvement depending on the level of risk (OECD, 2016b; OECD, 2009).

Building on the OECD's past work on risk management of droughts and floods in the agricultural sector (OECD, 2016b), this study analyses drought, flood and tropical storm risk management policies and producer strategies in a selected number of ASEAN countries with high risk exposure: Myanmar, the Philippines, Thailand and Viet Nam. The main objectives of this report are to:

- Explore the drought, flood and tropical storm risk exposure of selected ASEAN countries and its possible implications for their agricultural development and food security

- Assess current national policy approaches to address drought, flood and tropical storm risk in the agricultural sector and compare them with OECD recommendations and good practices

- Develop policy recommendations for ASEAN and its member countries to improve the management of drought, flood and tropical storm risk in agriculture.

The evidence gathered for this study came from a literature review of publications from international organisations offering guidelines and policy recommendations, project reports and scientific articles on the subject of droughts, floods and tropical storms in Southeast Asia, with a special focus on policy measures to tackle these risks in Myanmar, the Philippines, Thailand and Viet Nam. A data collection and validation exercise was undertaken through interviews with government stakeholders, international organisations and research institutes in Myanmar, Thailand and Viet Nam in February 2017. The Philippines was not included in the field visit because it was already well covered by the disaster risk management (DRM) literature reviewed.

The next section of this study gathers evidence from the literature on the risk exposure of the agricultural sector from the three weather-related disasters of droughts, floods and tropical storms in Southeast Asia. The following section reviews the policy measures and producer strategies in place in different ASEAN countries to mitigate and respond effectively to these weather-related disaster risks in light of the good practices identified in OECD countries, and with a focus on Myanmar, the Philippines, Thailand, and Viet Nam. In each section, the policies and practices implemented to address all three types of weather-related risks are described together rather than attempting to classify policies and practices according to a type of risk. Indeed, the analysis showed that the four countries studied had focused their efforts on responding to all of the three types of weather-related disasters in an indiscriminate way. Evidence of specific policy strategies targeting drought, floods or tropical storms was sparse.[2]

Box 2.2. Examples of policy measures to manage drought, flood and tropical storm risk in agriculture

While individual actions by farmers and their communities are essential, a co-ordinated set of public policies are needed to ensure the efficient allocation and prevention of risks by individuals, the market and governments (OECD, 2016b). OECD (2016b) developed a detailed list of actions to be implemented, at the individual and collective levels, to manage the risks linked to droughts and floods in the agricultural sector of OECD countries. Building on this framework and on other frameworks and policy documents, a selected list of *ex ante* and *ex post* measures are described below to illustrate the broad range of risk management tools that may be relevant in the context of the Association of Southeast Asian Nations (ASEAN) (ASEAN, 2016; Aye, 2014; Carter, 2008; CFE-DMHA, 2015; Chariyaphan, 2012; IFRC and UNDP, 2014; Myanmar, 2012; OECD, 2016a; OECD, 2009; Ortega, 2014; UNISDR and CRED, 2015). While a useful tool, these categories and classifications are not necessarily mutually exclusive and some policy tools can offer synergies across multiple policy objectives.

Examples of *ex ante* measures

Prevention and mitigation

Infrastructure: Hydrological infrastructure such as irrigation systems and dams are key to reducing weather-related disasters risk.

Information campaigns: Raising awareness about weather-related disasters risk (for example, through targeted extension services) can encourage shifts in on-farm strategies to reduce risk exposure.

Regulations: Water allocation rights that reflect water availability is a prerequisite to managing drought risk for the agricultural sector. Regulations prohibiting water pollution can also reduce water stress in the event of a drought. Moreover, spatial planning regulations to prohibit production in high-risk zones can reduce flood and tropical storm risks.

Incentive schemes: Some subsidies can be used to encourage risk-reducing investments (for instance, when market prices discourage farmers from investing in more efficient water systems such as drip irrigation, government subsidies to co-fund such investments can nudge farmers towards their adoption and thus reduce the impact of future droughts).

Insurance: With careful design, risk transfer tools such as insurance can support both prevention and preparedness. By raising awareness about risk exposure, certain insurance products can encourage risk-reducing behaviour on the farm. Insurance can also help farmers financially prepare for part of the remaining risk of weather-related disasters by transferring risk to the insurer.

Preparedness

Weather and hydrological information systems: Information systems help farmers adapt their production plan at the beginning of the cropping season. Early warning systems are also key to initiating emergency response processes in a timely manner.

Precautionary savings: Household savings are a useful financial tool to cope with liquidity constraints when weather-related disasters occur.

Examples of *ex post* measures

Response

Crisis management procedures: Clearly defined emergency response procedures can help mitigate the costs of water shortages or excesses.

Social protection: Cash payments (for instance, via disaster-linked cash transfer programmes) are sometimes provided to supplement revenue after severe natural disasters.

Recovery

Inputs or equipment: The provision of inputs and equipment can help farmers prepare for subsequent seasons. In some cases, payments are provided for the purchase of specific inputs or equipment.

Debt rescheduling: Loans schedules can be adjusted after a disaster occurs to ease liquidity constraints and the long-term impacts of income shocks.

Source: Australian Government Productivity Commission (2009); FAO (2015b); OECD (2009; 2016b; 2017a); UNISDR and CRED (2015); Sadoff et al. (2015).

Notes

[1] From now on, the phrase "weather-related disasters" will cover extreme droughts, floods and tropical storms that have an impact on human activity. Annex 2.A documents established definitions for the weather-related disasters covered by this report.

[2] Furthermore, the dichotomy of weather-related risks (droughts and floods on the one hand, and storms on the other), which can be relevant for OECD countries, did not match the situation for Southeast Asia. In the four countries studied, tropical storms and floods often come together following short-term meteorological events while droughts are separate long-term climatic events (Annex 2.A for definitions).

References

ADB (2013), *Myanmar: Agriculture, Natural Resources, and Environment Initial Sector Assessment, Strategy, and Road Map*, Asian Development Bank, Mandaluyong City, https://www.adb.org/documents/myanmar-agriculture-natural-resources-and-environment-initial-sector-assessment-strategy (accessed 28 February 2017).

Alano, E. and M. Lee (2016), "Natural disaster shocks and macroeconomic growth in Asia: evidence for typhoons and droughts", *ADB Economics Working Paper Series*, No. 503, Asian Development Bank, Mandaluyong City, https://papers.ssrn.com/sol3/papers.cfm?abstract_id=2894778 (accessed 27 February 2017).

ASEAN (2016), *ASEAN Disaster Recovery Reference Guide*, ASEAN Secretariat, Jakarta.

Australian Government Productivity Commission (2009), *Government Drought Support*. Productivity Commission Enquiry Report No. 46, Australian Government Productivity Commission, Melbourne, http://www.pc.gov.au/inquiries/completed/drought/report/drought-support.pdf (accessed 17 March 2016).

Aye, N.N. (2014), *Country report of Myanmar 2014*, Asian Disaster Reduction Center, Kobe, http://www.adrc.asia/countryreport/MMR/2014/FY2014A_MMR_CR.pdf (accessed 28 February 2017).

Carter, W.N. (2008), *Disaster management: a disaster manager's handbook*, Asian Development Bank, Mandaluyong City, https://www.adb.org/publications/disaster-management-disaster-managers-handbook (accessed 28 February 2017).

CFE-DMHA (2015), *ASEAN Disaster Management Reference Handbook*, Center for Excellence in Disaster Management & Humanitarian Assistance, Ford Island, Hawaii, http://reliefweb.int/sites/reliefweb.int/files/resources/ASEAN%20Disaster%20Management%20Reference%20Handbooks.pdf (accessed 1 March 2017).

Chariyaphan, R. (2012), *Thailand's country profile 2012*, Asian Disaster Reduction Center, Kobe, http://www.adrc.asia/countryreport/THA/2012/THA_CR2012B.pdf (accessed 28 February 2017).

EM-DAT (2017), *The EM-DAT Glossary*, http://www.emdat.be/Glossary (accessed 18 January 2017).

FAO (2016), Proceedings of the Workshop on Forests and Disasters in Southeast Asia, 24-25 August 2016, Antipolo City, Philippines, FAO, Bangkok.

FAO (2015a), *The Impact of Disasters on Agriculture and Food Security*, FAO, Rome, http://www.fao.org/resilience/resources/resources-detail/en/c/346258/ (accessed 29 March 2017).

FAO (2015b), *2015–2016 El Niño - Early action and response for agriculture, food security and nutrition - Update #4*, FAO, Rome, http://www.fao.org/resilience/resources/resources-detail/en/c/340660/ (accessed 29 March 2017).

FAO-MM (2017), *Myanmar at a glance*, http://www.fao.org/myanmar/fao-in-myanmar/myanmar/en/ (accessed 28 March 2017).

Field, C.B., et al. (eds.) (2014), *Climate Change 2014: Impacts, Adaptation, and Vulnerability. Part A: Global and Sectoral Aspects. Contribution of Working Group II to the Fifth Assessment Report of the Intergovernmental Panel on Climate Change*, Cambridge University Press, Cambridge and New York, http://www.ipcc.ch/report/ar5/wg2/ (accessed 29 March 2017).

IFRC and UNDP (2014), *Effective law and regulation for disaster risk reduction: a multi-country report*, IFRC and UNDP, New York, http://www.undp.org/content/undp/en/home/librarypage/crisis-prevention-and-recovery/effective-law---regulation-for-disaster-risk-reduction.html (accessed 28 February 2017).

Lesk, C., P. Rowhani and N. Ramankutty (2016), "Influence of extreme weather disasters on global crop production", *Nature*, Vol. 529, pp.84-87.

Lowder, S.K., J. Skoet and S. Singh (2014), "What do we really know about the number and distribution of farms and family farms worldwide?", Background paper for The State of Food and Agriculture 2014, *ESA Working Paper* No. 14-02, Food and Agriculture Organization of the United Nations, Rome.

Myanmar (2012), *Myanmar Action Plan on Disaster Risk Reduction (MAPDRR)*, The Republic of the Union of Myanmar, Nay Pyi Taw.

OECD (2017a), "Evaluation of farm programmes in the 2014 US Farm Bill: A review of the literature" [TAD/CA/APM/WP(2017)1/FINAL], OECD, Paris.

OECD (2017b), *OECD Recommendation on Disaster Risk Financing Strategies*, OECD Publishing, Paris, http://www.oecd.org/finance/oecd-recommendation-disaster-risk-financing-strategies.htm

OECD (2017c), *Water Risk Hotspots for Agriculture*, OECD Publishing, Paris, http://dx.doi.org/10.1787/9789264279551-en

OECD (2016a), *Financial Management of Flood Risk*, OECD Publishing, Paris, http://dx.doi.org/10.1787/9789264257689-en

OECD (2016b), *Mitigating Droughts and Floods in Agriculture: Policy Lessons and Approaches*, OECD Publishing, Paris, http://dx.doi.org/10.1787/9789264246744-en.

OECD (2014), *Recommendations on the Governance of Critical Risks,* OECD Publishing, Paris, http://www.oecd.org/gov/risk/recommendation-on-governance-of-critical-risks.htm

OECD (2011), *Managing Risk in Agriculture: Policy Assessment and Design,* OECD Publishing, Paris, http://dx.doi.org/10.1787/9789264116146-en.

OECD (2009), *Managing Risk in Agriculture: A Holistic Approach*, OECD Publishing, Paris, http://dx.doi.org/10.1787/9789264075313-en.

Ortega, A.P. (2014), *Philippines' country report 2014*, Asian Disaster Reduction Center, Kobe, http://www.adrc.asia/countryreport/PHL/2014/PHL_CR2014B.pdf (accessed 28 February 2017).

Raitzer, D.A., L.C.Y. Wong and J.N.G. Samson (2015), "Myanmar's agriculture sector: unlocking the potential for inclusive growth", *ADB economics working paper series*, No. 470, Asian Development Bank, Mandaluyong City.

Sadoff, C.W. et al. (2015), *Securing Water, Sustaining Growth: Report of the GWP/OECD Task Force on Water Security and Sustainable Growth*, University of Oxford, Oxford.

UN (2015), *Sendai Framework for Disaster Risk Reduction 2015 – 2030*, United Nations, New York,, http://www.preventionweb.net/files/43291_sendaiframeworkfordrren.pdf (accessed 29 March 2017).

UNFCCC (2015), *Agreement of the Conference of the Parties' 21st Session, Paris, 30 November – 11 December 2015*, UNFCCC, Geneva, http://unfccc.int/resource/docs/2015/cop21/eng/l09.pdf (accessed 29 March 2017).

UNISDR (2009), *Terminology on Disaster Risk Reduction*, UNISDR, Geneva, https://www.unisdr.org/we/inform/terminology.

UNISDR and CRED (2015), *The Human Cost of Weather Related Disasters 1995-2015*, UNISDR, Geneva, and CRED, Louvain, https://www.unisdr.org/we/inform/publications/46796 (accessed 29 March 2017).

University of Colorado (2016), *Floods and Droughts*, University of Colorado, Boulder, Colorado, http://www.colorado.edu/geography/class_homepages/geog_3511_s11/notes/Notes_11.pdf (accessed 18 January 2017).

World Bank (2016), World Development Indicators, http://databank.worldbank.org/data/ (accessed 21 April 2017).

Annex 2.A.

Definitions of weather-related hazards and extreme events

According to the UNISDR, a *hazard* is "a dangerous phenomenon, substance, human activity or condition that may cause loss of life, injury or other health impacts, property damage, loss of livelihoods and services, social and economic disruption, or environmental damage" (UNISDR, 2009). In its own commentary of this definition, the UNISDR explains that hazards that are relevant to disaster risk reduction are "hazards of natural origin and related environmental and technological hazards and risks." Such hazards arise from a variety of sources: geological (earthquakes), climatological (droughts), meteorological (storms), hydrological (floods), oceanic (storm surge), biological (epidemic), and technological (plane crashes) (UNISDR and CRED, 2015). Several hazards can sometimes act in combination; this is particularly the case for heavy storms, which can cause floods. Hazards can be described quantitatively by the likelihood of occurrence of different intensities for different areas, as determined from historical data or scientific analysis.

Extreme events are a special case of hazard and they can also be defined by the frequency of an observed variable. Extreme events are the most infrequent forms of hazard characterised by a probability of reoccurrence of 1 in 10 000 years (OECD, 2011). The IPCC defines an *extreme weather event* as the "occurrence of a weather variable above (or below) a threshold value near the upper (or lower) ends of the range of observed values and variables" (IPCC (2012), as quoted in OECD, 2016b).

Some weather-related disasters are cyclical and recurrent. Monsoon rainstorms and floods typically happen every year while the rainstorms and droughts caused by the *El Niño* southern oscillation occur every three to seven years. The term *El Niño* describes the abnormal warming of ocean water resulting from the oscillation of the ocean current in the South Pacific, usually accompanied by heavy rainfall in the coastal region of Peru and Chile, and reduction of rainfall in equatorial Africa and Australia (EM-DAT, 2017). With time, agricultural production systems have learned to adapt to annual weather cycles by making the best use of their cyclical conditions. However, the agricultural sector still struggles to deal with the effects of large *El Niño* southern oscillation events (FAO, 2015b). For example, the latest *El Niño* southern oscillation event (2015-2016) has affected 144 083 ha of Filipino farms, leading to losses in crops worth USD 70.8 million (FAO, 2015b). Another FAO study found that *El Niño* cycles were associated with more global agricultural area affected by extreme heat and drought conditions; these extreme conditions were particularly manifest in Asia (Alano and Lee, 2016).

Droughts

The main difference between meteorological and climatological disasters is their time scale. Climate-related extreme events are hazards "caused by long-lived, meso-to-macroscale atmospheric processes ranging from intra-seasonal to multi-decadal climate variability" (EM-DAT, 2017). EM-DAT classifies droughts as climatological disasters because they occur over a long period of months to years.

The geography course at the University of Colorado defines drought as "an extended period (months, years) that a region experiences a deficiency in water supply, generally because of reduced precipitation" (University of Colorado, 2016). It goes on to characterise three basic sequences of drought and associated impacts:

- Meteorological drought: "deficit in precipitation, often (not always) accompanied by above average temperatures, high winds, low humidity and high solar radiation."

- Agricultural drought: "continued precipitation deficit, leading to a soil water deficit, hindering agriculture and natural plant growth."

- Hydrological drought: "the precipitation deficit continues, and stream discharge, lake wetland and reservoir levels drop, with impacts on wildlife habitat."

This definition of drought starts to introduce the complexity of weather phenomena, which are often associated or lead to one another. FAO (2015a) uses a two-step definition where meteorological drought can lead to environmental drought, which combines agricultural and hydrological drought from the University of Colorado definition. OECD (2016b) defines drought as a "temporary decrease of water availability in a given water system, caused by prolonged deviations from average levels precipitation" so also uses the two-step definition where meteorological drought leads to environmental drought. The rest of the OECD (2016b) report on floods and droughts is in line with the EM-DAT definition of drought in that droughts are made more complex by the possible interaction of human activities and water supply demands with the physical lack of water. EM-DAT (2017) thus mentions that in parallel with the conceptual definition of a drought, operational definitions specifying the level of precipitation that qualifies as a drought vary according to locality, climate and environmental sector.

Floods

According to the open-access geography course of the University of Colorado, "floods occur when a drainage basin experiences an unusually intense or prolonged water input. Flood is usually viewed as an event in which the streamflow exceeds the channel capacity, resulting in overland flow [...], but the term is also often applied more generally to unusually high discharge events" (University of Colorado, 2016).

OECD relates floods to the change in water level, using the following definition: "rise, usually brief, in the water level of a stream or water body to a peak from which the water level recedes at a slower rate" (OECD, 2016a). The main types of flood risk that affect agriculture are: riverine floods, flash floods and coastal floods linked to storm surges (EM-DAT, 2017).

- A riverine flood is an overflow of water from a stream channel onto normally dry land in the floodplain.

- A flash flood is a rapid inland flood due to intense rainfall. A flash flood describes sudden flooding with short duration. In sloped terrain the water flows rapidly with a high destruction potential. In flat terrain the rain water cannot infiltrate into the ground or run off (due to small slope) as quickly as it falls. Flash floods typically are associated with thunder storms. A flash flood can occur at virtually any place.
- Coastal floods are characterised by higher-than-normal water levels along the coast caused by tidal changes or thunder storms that result in flooding, which can last from days to weeks.

Typhoon and cyclone tropical storms

Meteorological disasters are "events caused by short-lived or small-to-mesoscale atmospheric processes" lasting from minutes to days (EM-DAT, 2017). The meteorological disasters that are most likely to impact on agriculture are storms (tropical storms, thunder storms, hail storms, strong winds) and extreme temperatures (freeze, frost, heat waves and cold spells).

In tropical Southeast Asia, the most common meteorological disaster is a tropical storm. A tropical storm originates over tropical or subtropical waters. It is characterised by a warm-core, non-frontal synoptic-scale cyclone with a low pressure centre, spiral rain bands and strong winds. Depending on their location, tropical storms are referred to as hurricanes (Atlantic, North-East Pacific), cyclones (South Pacific and Indian Ocean, including Myanmar), or typhoons (North-West Pacific, and thus the other Southeast Asian countries covered by this report). The observed increase in sea surface temperatures is likely to lead to higher frequencies of tropical storms and heightened storm surges. The concomitant rise of global sea levels will also lead to higher storm surges (Alano and Lee, 2016). The English term "typhoon" derives from the Cantonese "taiphoong", or big wind.

Chapter 3. High exposure to weather-related risks in ASEAN countries

This chapter provides detailed evidence of the high exposure to weather-related risks in the agricultural sector in Myanmar, the Philippines, Thailand, and Viet Nam.

3.1. The increasing exposure of ASEAN countries to agricultural water risks

Southeast Asia is one of the most exposed regions to weather-related disasters (Global Climate Risk Index, 2017). In many of these countries, the agricultural sector is bearing the brunt of the economic impact (FAO, 2015a). Tropical storms can provoke major damage to agriculture by destroying agriculture-related infrastructure or crops in fields. Likewise, flash floods can destroy crops in the field or sweep away livestock. Prolonged flooding can also end up drowning annual and perennial crops. Droughts can have a severe impact on a country's crop production but their impacts are also important on the livestock sector, which suffers nutritional impacts from lack of fodder (FAO, 2015a).

Due to their extended land mass covering several ecological and climatic tropical zones from north to south, Myanmar, the Philippines, Thailand and Viet Nam can all find themselves in a situation where they have to cope at once with both excess water and lack of water, or in short sequence, in different parts of the country.

Climate change is likely to exacerbate the extreme nature of weather-related disasters, with major impact on the agricultural sector in ASEAN. Because of climate change, rainfall patterns will change significantly up to 2040. Tropical storms are likely to intensify by 10 to 20 percentage points from current levels up to 2050. Annual average temperatures in the Philippines, Thailand and Viet Nam are projected to increase by 4.8°C up to 2100, and the global average sea level could rise by 70 cm because of climate change (Cruz et al., 2007; ADB, 2009). All these local weather effects from climate change are likely to be compounded by the regular *El Niño* cycles affecting the region (Alano and Lee, 2016; FAO, 2015b). As a result, short-term crop failures and long-term production declines are more likely to occur. Droughts are expected to become more intense and more frequent. Given that droughts currently have the greatest negative effect of all weather-related disasters on ASEAN annual agricultural production, these developments could have a significant negative impact on regional production and farm household incomes. This worrying scenario is analysed in greater detail for the Philippines in *Agricultural policies in the Philippines* (OECD, 2017a).

This chapter analyses the risk exposure of the selected ASEAN case countries – Myanmar, the Philippines, Thailand and Viet Nam – to these weather-related events, and the damages and losses suffered in recent years. Table 3.1 presents a summary of data from various sources for a snapshot of the magnitude of the risk exposure of these four countries to weather-related disasters and damages caused to agriculture based on frequency data from the past two decades. It shows that three of the four countries studied have been hit by several disastrous floods every year on average in the past two decades, while Myanmar has had one disastrous flood nearly every year on average during the same period. The Philippines and Viet Nam have also witnessed an especially high frequency of disastrous typhoons in the last two decades. Furthermore, the damages and losses to agriculture from the most recent milestone weather-related disaster in the four countries have been important. In particular, the 2008 Nargis cyclone in Myanmar and the 2011 monsoon floods in Thailand had a significant impact on national rice paddy yields.[1]

3.2. Agriculture's exposure and vulnerability to weather-related disasters in Myanmar, the Philippines, Thailand, and Viet Nam

Myanmar suffers from high exposure to flood, cyclone and drought risk, which poses a key threat to the agricultural sector (Raitzer et al., 2015). As a heavy rainfall country, flooding is common (ADRC Myanmar, 2017; FAO, 2009). With its long coastline on the eastern side of the Bay of Bengal, cyclones are also frequent events. "Annually, about ten tropical cyclones and depressions form in the Bay of Bengal and out of them almost five become severe cyclones and cross the coasts of […] Myanmar" (Aye, 2014). Cyclone Nargis in 2008 was particularly severe. These events have negatively impacted the agricultural sector (Table 2.1).

The Philippines' agricultural sector is also highly exposed to flood, typhoon and drought risk. Situated on the Pacific's typhoon belt, the Philippines experiences 20 typhoons per year on average, approximately five of which are severe in a given year (ADRC Philippines, 2017). Most recently, Typhoons Haiyan (in 2013) and Ondoy and Pepeng (in 2009) were particularly damaging (FAO, 2015c; World Bank, 2011). "Damage to agricultural production in the Philippines [from such events] is substantial. Between 2006 and 2013, the FAO (2015a) estimates that total damage and loss in the agriculture sector amounted to USD 3.8 billion, arising from 78 natural disasters: two droughts, 24 floods, 50 tropical cyclones or tropical storms, one earthquake and one volcanic eruption. The majority of damage and losses (USD 3.1 billion) occurred in the crop subsector, with over 6 million ha of crops affected (FAO, 2015a)" (OECD, 2017a). High risk exposure is contributing to lower productivity growth in the Philippines' agricultural sector relative to other countries in the region (OECD, 2017a). "Land use and yields are especially affected by *El Niño* cycles which are hard to predict and typically last between 14-22 months. *La Niña* cycles have the opposite effect: conditions are wetter than normal, intensifying monsoon rains and increasing the risk of flooding and mudslides" (OECD, 2017a).

Three agricultural producing regions in Thailand are particularly prone to weather-related disasters. The northeast plateau suffers from regular droughts and flash floods. The central river basin is subject to yearly monsoon floods. The hilly south is hit by typhoons and flooding. Rice is dominant in the country's agriculture, making Thailand a prominent world producer and exporter (Chariyaphan, 2012). It is estimated that over 10% of the country's rice area suffered from natural disasters between 2005 and 2014 (Rattanayod, 2016). Droughts and floods are particularly damaging for the agricultural sector (ADRC Thailand, 2017). A growing trend towards mono-cropping also raises concerns about resilience. In particular, rubber production in the south is prone to erosion in the flood season. Despite the high sensitivity of single-crop rubber plantations to soil erosion, this production system has been commercially successful in the southern region for several years, and has been promoted in other regions like the east and northeastern regions without careful consideration of the environmental risks or of the economic risks from changes in world market prices. Moreover, natural forest in the north (particularly in Nan Province) is being converted to maize production, which is not a particularly resilient crop for this region (EEPSEA, 2017). The 2011 monsoon flood was an extreme disaster for the country and badly affected the agricultural sector (Table 3.1). Drought in the north and northeast is also a regular event. In 2010, Thailand experienced USD 450 million in crop damages due to severe drought (OECD, 2017b).

Table 3.1. The exposure of the agricultural sector to weather-related disasters in selected ASEAN countries

	Myanmar	Philippines	Thailand	Viet Nam
Frequency of disastrous drought[1]	0%	20%	20%	25%
Frequency of disastrous flood[2]	95%	100% (on average 4.7 disastrous floods per year)	100% (on average 2.65 disastrous floods per year)	100% (on average 3.05 disastrous floods per year)
Frequency of disastrous storm[3]	35%	100% (on average 7.65 disastrous storms per year)	75%	100% (on average 2.95 disastrous storms per year)
Recent milestone weather-related disaster	Cyclone Nargis 2 May 2008	Typhoon Haiyan 8 November 2013	Monsoon flood 25 July 2011 to 16 January 2012	Mekong River Delta, South Central and Central Highlands drought 2015-2017
Annual paddy yield around milestone weather-related disaster[4]	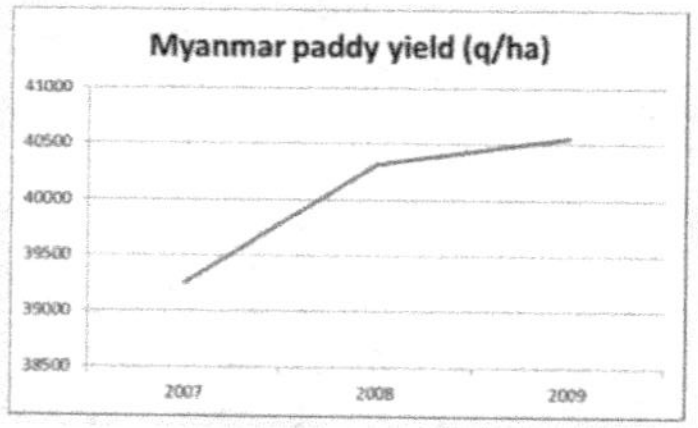 Myanmar paddy yield (q/ha)	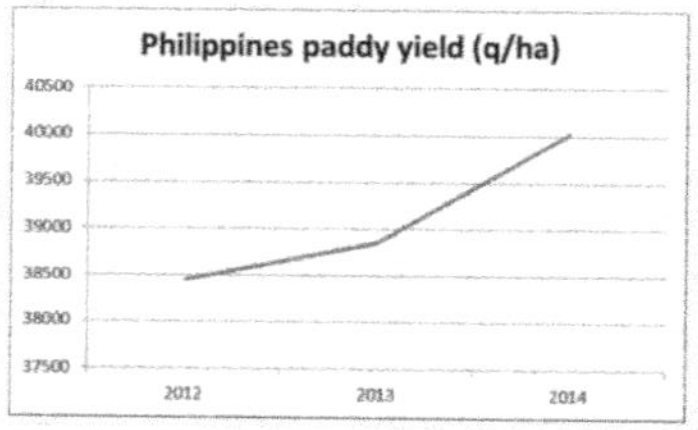 Philippines paddy yield (q/ha)	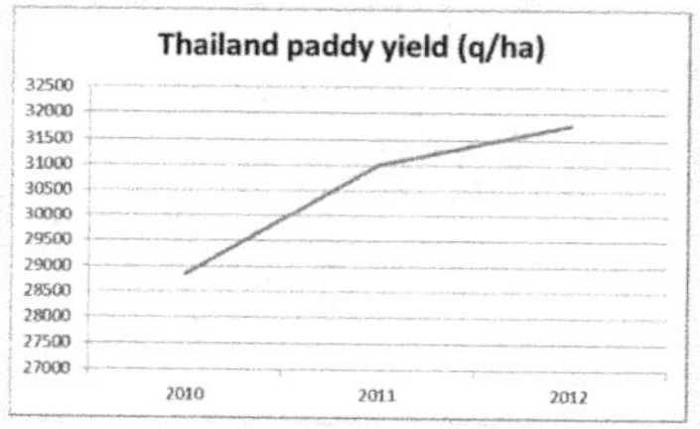 Thailand paddy yield (q/ha)	Not available
Estimated damages and losses to the agricultural sector from milestone weather-related disaster	783 220 ha (63%) of paddy fields submerged; 707 500 t of stored paddy and milled rice destroyed; 85% of seed stocks lost; some 3 000 power tillers and thousands of tilling equipment lost; 37% of orchard crops and 70% of backyard gardening destroyed; 50% of buffaloes and 25% of cattle lost in the 11 most affected townships alone (FAO, 2009)	USD 700 mln in damages to the agricultural sector; 600 000 ha of farmland (crops, orchards, plantations) affected; 44 mln coconut trees damaged or destroyed; 1.1 mln tonnes of crops lost (FAO, 2015c)	USD 1.1 tln in damages and losses to food crops; USD 138 mln in damages and losses to perennial crops; USD 104 mln in damages and losses to livestock (Ministry of Finance and World Bank, 2012)	Local rice yield dropped by 99% and by 76% for vegetables in the Mekong delta; local maize yield dropped by 92% in the South Central region; local coffee yield dropped by 81 percentage points and by 80% for cashew nuts in the Central Highlands; 45% of chicken lost in the Mekong delta; 50% of cattle and buffalo reported sick (FAO, 2016)[5]

Notes: [1]: Probability of a year having a disastrous drought. Calculations based on EM-DAT data 1997-2016.

[2]: Probability of a year having a disastrous flood. Calculations based on EM-DAT data 1997-2016.

[3]: Probability of a year having a disastrous storm. Calculations based on EM-DAT data 1997-2016.

[4]: *Source*: FAOSTAT http://www.fao.org/faostat/en/#home. The short-term trend using national data is not the best indicator to represent the impact of the disasters at the local level. These trends do not imply a direct relationship between the weather-related disaster mentioned in this table and national rice yields, which can be impacted by several other parameters. In particular, typhoon Haiyan struck in the Central Visayas of the Philippines without affecting the main rice production areas in Luzon so it seems not to have had an impact on the national rice yield. Haiyan was more damaging to other crops and livestock in the affected areas (FAO, 2015c).

[5]: This FAO agricultural post-disaster needs assessment was based on focus group discussions in highly affected villages so the numbers are not representative for the overall region or country. More robust post-disaster needs assessments are needed to evaluate the damages and losses from the recent drought.

Viet Nam's location on the tropical typhoon belt makes it a particularly hazard-prone country. Weather-related disasters together with landslides were responsible for losses equal to 1.5% of annual GDP between 2001 and 2010. Despite strong investment in poverty reduction over the past 20 years, the country's poorer populations are now concentrated mostly in rural areas and are highly dependent on climate-sensitive livelihoods like agriculture. With 80% of Vietnamese farmers growing rice on 45% of the country's agricultural land, weather-related disasters can have a very strong impact on the country's food security and could destabilise international markets given the country's leading export status. Production losses due to these risks could total 9.1 million tonnes annually by 2050 (USAID, 2017). Other food and cash crops are also vulnerable to weather-related disasters. Sea level rise due to climate change and land subsidence due to increased groundwater pumping for irrigation (Minderhoud et al., 2017) have added an additional threat with higher risk of sea water intrusion and increased soil salinisation in the Mekong Delta, which produces 13% of the world's rice. Although water is abundant in the country, it is not evenly distributed. The recent 2016 drought in the Mekong River basin was the worst recorded in the past 90 years (FAO, 2016).

Given that the agricultural sector in these four Southeast Asian countries has been regularly subject to droughts, floods and tropical storms, farmers, their communities and governments have adopted practices and institutions that have slowly contributed to increase their resilience. The next section elaborates on actions taken by governments to address these risks.

Note

[1] It is important to note that the data on frequency of disasters presented in Table 3.1 provide an aggregated national snapshot. However, the risks linked to weather-related disasters are – in reality – more geographically specific, with certain areas affected by floods and tropical storms every year and other areas more prone to drought, as briefly presented in the country-specific paragraphs that follow

References

ADB (2009), *The Economics of Climate Change in Southeast Asia: A Regional Review*, Asian Development Bank, Mandaluyong City.

ADRC Myanmar (2017), Information on Disaster Risk Reduction of the Member Countries Myanmar, http://www.adrc.asia/nationinformation.php?NationCode=104&Lang=en&Mode=country (accessed 29 March 2017).

ADRC Philippines (2017), Information on Disaster Risk Reduction of the Member Countries Philippines, http://www.adrc.asia/nationinformation.php?NationCode=608&Lang=en&NationNum=14 (accessed 29 March 2017).

Alano, E. and M. Lee (2016), "Natural disaster shocks and macroeconomic growth in Asia: evidence for typhoons and droughts", *ADB Economics Working Paper Series*, No. 503, Asian Development Bank, Mandaluyong City, https://papers.ssrn.com/sol3/papers.cfm?abstract_id=2894778 (accessed 27 February 2017).

Aye, N.N. (2014), *Country report of Myanmar 2014*, Asian Disaster Reduction Center, Kobe, http://www.adrc.asia/countryreport/MMR/2014/FY2014A_MMR_CR.pdf (accessed 28 February 2017).

Chariyaphan, R. (2012), *Thailand's country profile 2012*, Asian Disaster Reduction Center, Kobe, http://www.adrc.asia/countryreport/THA/2012/THA_CR2012B.pdf (accessed 28 February 2017).

Cruz, R.V. et al. (2007), "Asia" in *Climate Change 2007: Impacts, Adaptation and Vulnerability. Contribution of Working Group II to the Fourth Assessment Report of the Intergovernmental Panel on Climate Change,* M.L. Parry, O.F. Canziani, J.P. Palutikof, P.J. van der Linden and C.E. Hanson (eds.), Cambridge University Press, Cambridge, pp. 469-506.

EEPSEA (2017) Interview with Economy and Environment Program for Southeast Asia (EEPSEA), Bangkok, 7 February 2017.

EM-DAT (2017), The EM-DAT Glossary, http://www.emdat.be/Glossary (accessed 18 January 2017).

FAO (2016), *"El Niño" event in Viet Nam: agriculture, food security and livelihood needs assessment in response to drought and salt water intrusion*, FAO, Hanoi, http://www.fao.org/3/a-i6020e.pdf (accessed 3 March 2017).

FAO (2015a), *The Impact of Disasters on Agriculture and Food Security*, FAO, Rome, http://www.fao.org/resilience/resources/resources-detail/en/c/346258/ (accessed 29 March 2017).

FAO (2015b), *2015–2016 El Niño - Early action and response for agriculture, food security and nutrition - Update #4*, FAO, Rome, http://www.fao.org/resilience/resources/resources-detail/en/c/340660/ (accessed 29 March 2017).

FAO (2015c), *Typhoon Haiyan – Portraits of Resilience*, FAO, Rome, http://www.fao.org/resilience/resources/resources-detail/en/c/357721/ (accessed 29 March 2017).

FAO (2009), *Post-Nargis Recovery and Rehabilitation Programme*, FAO, Yangon, http://www.fao.org/fileadmin/user_upload/emergencies/docs/FAO_Myanmar__Post_Nargis_Recovery_and_Rehabilitation_Programme_Strategy.pdf (accessed 29 March 2017).

Global Climate Risk Index (2017), https://germanwatch.org/en/cri (accessed 16 March 2017).

Minderhoud, P.S.J. et al. (2017), "Subsidence in the Mekong Delta; contribution of groundwater exploitation to total subsidence"," *Geophysical Research Abstracts*, Vol. 19, EGU2017-12032.

Ministry of Finance and World Bank (2012), *Thailand Flooding 2554 Rapid Assessment for Resilient Recovery and Reconstruction Planning*, Thai Ministry of Finance and World Bank, Bangkok, http://www.undp.org/content/dam/thailand/docs/UNDP_RRR_THFloods.pdf (accessed 29 March 2017).

OECD (2017a), *Agricultural Policies in the Philippines*, OECD Publishing, Paris, http://dx.doi.org/10.1787/9789264269088-en.

OECD (2017b), *Building Food Security and Managing Risk in Southeast Asia*, OECD Publishing, Paris, http://dx.doi.org/10.1787/9789264272392-en

Raitzer, D.A., L.C.Y. Wong and J.N.G. Samson (2015), "Myanmar's agriculture sector: unlocking the potential for inclusive growth", *ADB economics working paper series*, No. 470, Asian Development Bank, Mandaluyong City.

Rattanayod, B. (2016), "Sovereign Risk Transfer Solutions Thailand", Presentation to the GiZ Executive Consultation and Capacity Building Seminar for Asian Government Representatives, 17-19 October 2016, Jakarta, Indonesia, GiZ, Bangkok, Thailand, https://jakarta-2016.climate-risk-insurance.org/ (accessed 5 April 2017)

USAID (2017), "Climate change risk profile Vietnam", USAID Fact Sheet, USAID, Hanoi, https://www.climatelinks.org/resources/climate-change-risk-profile-vietnam (accessed 29 March 2017).

World Bank (2011), *Typhoon Ondoy and Pepeng post disaster needs assessment*, World Bank, Washington, DC, http://documents.worldbank.org/curated/en/551271468296967206/Executive-summary (accessed 29 March 2017).

Chapter 4. Review of disaster risk management policies for agriculture in selected ASEAN countries

This chapter provides an overview of the legal basis and institutions for disaster risk management and its relevant stakeholders in the four case study countries covered by this study: Myanmar, the Philippines, Thailand and Viet Nam. Prevention and mitigation measures to reduce risk exposure are reviewed, followed by an evaluation of strategies to anticipate risks that cannot be eliminated. Response measures to reduce the effects of a disaster in the short-term are then analysed. Finally, recovery measures to help the agricultural sector bounce back are reviewed.

A broad range of *ex ante* and *ex post* measures are employed by governments in the four ASEAN countries under study (Myanmar, the Philippines, Thailand and Viet Nam) to cope with the risk of droughts, floods and tropical storms, and to strengthen agricultural resilience. These policy measures complement and strengthen the arrangements already taken by farmers, their communities and business partners to face these risks. Some countries are more advanced than others in certain policy areas (ASEAN, 2016; CFE-DMHA, 2015), providing a useful opportunity for sharing good practices within the region and more broadly.

4.1. Legal frameworks and institutions for disaster risk management in Southeast Asian countries

As recommended by international organisations supporting disaster risk management in Asian developing countries (ASEAN, 2016; Carter, 2008; CFE-DMHA, 2015; IFRC and UNDP, 2014; UN, 2015) and as highlighted by recent OECD guidelines on DRM (OECD, 2014b and 2017d), most Southeast Asian countries use existing government structures as the basis for managing disasters. These structures are supported by specialised government agencies when necessary. These *ad hoc* organisations can take the form of a National Disaster Council to take policy decisions, a National Disaster Management Office for the day-to-day implementation of disaster risk reduction (DRR), and other appropriate agencies and committees at lower levels of government in the areas at risk or already affected (Carter, 2008).

The activities of these commissions and agencies contribute to developing a national disaster management policy covering the following:

- Accurate definition of the disaster threat

- Identification of the likely effects of the threat

- Assessment of the resources available to manage the threat

- Organisational arrangements needed to prepare for, respond to and recover from disasters

- Definition of the linkages between a national disaster management policy and other national policies on development and environmental protection.

The policy framework in ASEAN countries, however, is still relatively weak regarding strategies to finance disaster risk prevention, preparedness, response and recovery (UNDP, 2017a). As a template that could help fill this gap, the Recommendation on Disaster Risk Financing Strategies adopted by OECD in February 2017 provides a set of high-level recommendations for designing a strategy that addresses the financial impact of disasters on individuals, businesses, and sub-national levels of governments, as well as the implication for public finances (OECD, 2017d). In particular, it highlights the need to examine the relative cost-effectiveness of different approaches to managing the financial impacts of disasters (i.e. prevention, preparedness, response, recovery). The OECD Recommendation on the Governance of Critical Risks also emphasises an integrative risk management approach, including risk prevention and mitigation (OECD, 2014b).

The legal framework for DRM and key institutions in the four case countries are as follows.

In Myanmar, the Disaster Management Law of 2013 establishes the different government bodies that co-ordinate DRM. The law is developed to be in line with the Hyogo Framework for Action (2005-2015) and to comply with the ASEAN Agreement on Disaster Management and Emergency Response (AADMER). The law includes provisions for the establishment of disaster management bodies, and their duties and responsibilities for all phases of disaster, as well as the establishment of a disaster management fund at national and at regional or state level. The law also provides the guidance to carry out DRR measures in line with the overall development plans of the country (ADRC Myanmar, 2017; Aye, 2014).

A National Disaster Management Agency chaired by the Ministry of Social Welfare, Relief and Resettlement co-ordinates the government's activities in DRM across all departments. The work of the Agency is implemented within the Working Committee on National Disaster Management and Relief, which is chaired by the Ministry of Construction, with the Ministry of Agriculture, Livestock and Irrigation as Associate-Secretary. National DRM plans serve as the backbone for the elaboration of disaster prevention and preparedness plans in states, regions and districts.

The Myanmar Action Plan on Disaster Risk Reduction (MAPDRR) provides a framework for multi-stakeholder engagement on DRR (Myanmar, 2012). It was prepared in substantial consultation with various stakeholders. MAPDRR's goal is "to make Myanmar safer and more resilient against natural hazards, thus protecting lives, livelihood and development gains". MAPDRR identifies 65 projects that need to be implemented to meet the government's commitments to the Hyogo Framework for Action and the AADMER. MAPDRR was launched in 2012 and some projects are still on-going. The lessons learned from its implementation were discussed between government services and the international support community at a dedicated workshop in April 2016 (Myanmar, 2016). An Emergency Operation Centre has been set up and is now operational in Nay Pyi Taw to tackle the challenges identified by the MAPDRR of poor co-ordination amongst and between agencies. The MAPDRR has also developed action plans at the local level for rapid local government response in case of floods and cyclones.

The Myanmar government generally welcomes the assistance of the international community in improving its DRM (FAO-MM, 2017; World Bank, 2017). However, there is still little co-ordination between international development partners. Furthermore, the international partners are not all present at discussions on the MAPDRR (FAO-MM, 2017). Many of the activities described in this report have been funded by bilateral development agencies from Japan, New Zealand, and by international development partners like the UN and the Asian Development Bank.

The 2010 Philippines Disaster Risk Reduction and Management Act of 2010 sets the legal basis for a "holistic, comprehensive, integrated and proactive" DRM, including disasters caused by climate change. The Act is the foundation of the National Disaster Risk Reduction and Management Plan of 2010 to implement DRM across the country (ADRC Philippines, 2017; Ortega, 2014). The National Disaster Risk Reduction and Management Council (NDRRMC) is the government body empowered to co-ordinate, integrate, supervise, monitor and evaluate DRM policy-making and activities. It can call upon other non-government and civil society institutions to assist its activities. The Office of Civil Defence is its implementing arm.

Presidential Executive Order 888 of 2010 set up the Strategic National Action Plan on DRR for 2009-19 as a road map indicating the vision and strategic objectives on DRR for the country over the following ten years based on an assessment of disaster risks, gaps analysis, and activities proposed by the Hyogo Framework for Action that local stakeholders thought achievable by the country. Due to a lack of resources, most of the priority projects of the Strategic National Action Plan have not yet been fully realised (Ortega, 2014). The NDRRMC adopted the 2011 National Disaster Risk Reduction and Management Framework setting the vision for disaster-resilient Filipino communities, indicating a paradigm shift towards proactive and preventive approaches to DRM. A devolution process mandates local government units to organise Provincial, City and Municipal DRR and Management Councils. However, local government units are in general technically ill equipped, face budget and manpower constraints, all of which make them unable to address the expected policy outcomes devolved to them. Furthermore, some local councils have other priorities which affects their capability for extension and DRM. This warrants a second look at the rationale for devolution (SEARCA, 2017; World Bank, 2011).

The Philippine government has welcomed international assistance to cope with large-scale weather-related disasters. The frequent occurrence of these disasters has led to established channels for co-operation and aid, which are strengthened and improved each time they are called upon (ADRC Philippines, 2017; FAO, 2016b; FAO, 2015c; OECD, 2017a; Ortega, 2014; UNDP, 2017a; UNISDR, 2013a; World Bank, 2011).

In Thailand, the Disaster Prevention and Mitigation Act of 2007 nominates the Department of Disaster Prevention and Mitigation (DDPM), under the Ministry of Interior, as the government institution co-ordinating national DRM activities. The act also authorises local government to co-ordinate local DRM activities. The National Disaster Prevention and Mitigation Committee chaired by the Prime Minister is the interdepartmental body in charge of policy making on DRM. A National Civil Defence Committee is the implementation arm of the DDPM to manage disasters at national level. In line with the Hyogo Framework for Action, Thailand has developed a Strategic national Action Plan for DRR (ADRC Thailand, 2017; Chariyaphan, 2012). The country has also elaborated a National Disaster Prevention and Mitigation Plan in 2015 in accordance with the UN Sendai Framework (FAO, 2016b). Thailand has allotted a specific budget for DRM. Its first component funds DRR activities within the Strategic National Action Plan and contributes funding to Post-Disaster Needs Assessments (PDNAs). Its second component funds the activities implemented by other line ministries when they respond to disasters affecting economic activities falling under their mandate (UNDP, 2017a). The Ministry of Agriculture and Cooperatives has prepared a disaster preparedness plan to mitigate the effects of the 2016-17 drought, together with a rehabilitation plan for the flood victims in the Southern Region that had occurred simultaneously. Despite the co-ordinating activities of the DDPM, collaboration between ministries involved in DRM at the national level is not optimal. Likewise, the line departments at the provincial level are often confined to their silos and cross-sectoral co-ordination is less than optimal. Most decisions on national and even local DRM requiring interministerial collaboration are sent back to the Ministry of the Interior or the Prime Minister's Office (UNDP-TH, 2017).

The Thai government does not usually request international support to cope with weather-related disasters, preferring to rely on its own resources or to make use of regional mechanisms co-ordinated by the ASEAN Secretariat. Nevertheless, UNDP has supported the Thai government in the drafting of the Disaster Prevention and Mitigation Act, the Ministry of Agriculture and Cooperatives has expressed interest to receive technical assistance from UNDP to implement the PDNA method in the agricultural sector, and the Ministry of Foreign Affairs has requested the UN Country Team to develop and conduct a general PDNA training course for DRM focal points in all ministries (UNDP, 2017a; UNDP-TH, 2017).

In Viet Nam, the 1990 Decree No. 168-HDBT created the Central Committee of Storm and Flood Control, an interministerial institution that co-ordinates DRR in the country. Its secretariat is held by the Department of Dike Management and Flood Control of the Ministry of Agriculture and Rural Development (MARD). The interministerial Central Committee is translated at the provincial level as a Provincial Committee for Flood and Storm Control, which co-ordinates local emergency activities and makes use of the technical assistance from the various government departments under its local authority. The country elaborated a First National Strategy and Action Plan for Mitigating Water Disaster in 1994 through a national consultation process. The first strategy put special emphasis on reducing the country's vulnerability to weather-related disasters and to improve its capacity to cope with them. The Second Strategic Action Plan (2001-20) went further in setting up strategies for disaster mitigation and management (ADRC Viet Nam, 2017).

Since 2014, the government has started collecting a small contribution from all Vietnamese workers and businesses to constitute a National Disaster Prevention Fund. The fund is designed to help cover the costs provinces face to recover from disasters, and to provide water, food, and cash transfers to affected households (MONRE, 2014). The fund is currently worth USD 22 million (MARD, 2017b), but is still far away from the estimated USD 660 million the Vietnamese economy needs to repair the damages caused by weather-related disasters each year.

The Vietnamese government had to request international assistance in 2016 because its DRR systems had been overwhelmed by the onslaught of repeated disasters that year (winter frost in the North, typhoons in the Central provinces and Red River Valley, drought in the Mekong River Delta, the Southeast and Central Highlands), which followed an already disastrous 2015 (Australian Embassy to Viet Nam, 2017; US Embassy to Viet Nam, 2017). Generally speaking, Viet Nam is keen to learn about policy and technical innovations from other countries, which could be imported and adapted to help the country mitigate and cope with weather-related disasters.

While the response and preparedness components of DRM are quite similar across all sectors of the economy, the prevention and preparedness measures are quite specific in the agricultural sector. The ministry in charge of agriculture and its related institutions in the ASEAN countries studied play an important role in helping the agricultural sector at large mitigate risk from, and prepare itself against weather-related disasters, as will be detailed below.

As mentioned above, individual and local community actions are important elements to prevent, prepare for, and respond to a weather-related disaster. In the ASEAN countries where these disasters are relatively regular, there is a long tradition of local knowledge on how best to cope with these extreme events (Box 4.1). Many of the traditional mixed-crop and livestock farm production systems in the tropical areas of Southeast Asia have developed historically to spread food security risk across different enterprises given the local agro-ecological environment (Losch, 1996). The introduction by the Green Revolution of more productive hybrid seeds and yield-enhancing agricultural inputs has led to impressive progress in their food security status, although often to the detriment of integrated risk management by farm households (Conway, 2012). By encouraging farmers towards the production of a limited number of high-yielding or economically profitable commodities, some public policies have contributed to the loss of some of the traditional risk management knowledge and farm practices deposited through historical trial and error, and shared within farming communities (EEPSEA, 2017).

Associated with modern information sharing tools and best practice advice derived from latest research and innovation results, traditional knowledge can be a starting point and complementary component to the development of effective community-based disaster risk reduction and management (Seng, 2016). Community-based activities are a first step to institutionalising DRM among the people at a local level, so it is important that national and local government planning and action make use and reinforce this traditional know-how by associating it to state-of-the-art agronomic innovation and modern information and communication technologies like mobile phones that have become nearly ubiquitous in ASEAN rural areas. In instances where traditional and local knowledge are not enough to cope with weather-related disasters because of market failures, government intervention through appropriate policy instruments is still required.

To complement the traditional and indigenous knowledge for mitigating weather-related disasters, a structured combination of complementary policy instruments is needed (OECD, 2016c). Each instrument should strive to target a specific market or behavioural failure. According to former OECD work on agricultural risk management (OECD, 2009), government intervention in risk management should be limited to areas beyond the capacity of farmers alone or with the market-based instruments available to them (Box 4.2).

Box 4.1. Anecdotal evidence of indigenous traditional and local knowledge to cope with disasters

"Indigenous knowledge refers to the methods and practices developed by a group of people from an advanced understanding of the local environment, which has formed over numerous generations of habitation. This knowledge contains several other important characteristics which distinguish it from other types of knowledge. These include originating within the community, maintaining a non-formal means of dissemination, collectively owned, developed over several generations and subject to adaptation, and imbedded in a community's way of life as a means of survival" (UNISDR, 2008).

Indigenous knowledge is valuable to disaster risk reduction (DRR) policies in four ways:

- Various specific indigenous practices and strategies can be easily transferred and adapted to other communities facing similar contexts
- Incorporating indigenous knowledge in existing practices and policies fosters the participation and leadership of local communities and their members in DRR activities
- Traditional knowledge provides invaluable information on the local context for disaster risk management (DRM)
- The informal way in which indigenous knowledge is transferred provides a successful model for training on DRR.

Some examples of how local communities in countries of the Association of Southeast Asian Nations (ASEAN) have used indigenous knowledge to cope with weather-related disasters are enlightening for enriching DRM policies with already existing practices.

The *kanungkong* is a traditional communication device from Northern Luzon, Philippines, using bamboo poles. Dagupan City has adapted the *kanungkong* and incorporated it into its early warning system. The local authorities and communities agreed on a code of rhythm (number of strikes and time intervals) and sound to correspond to specific actions to be taken by individuals in preparedness for a disaster. The sound code corresponds to international disaster warning colour standards. One *kanungkong* positioned every five houses relays the warning along the river banks.

In the Batanes islands north of the Philippines, strong winds have shaped local livelihoods for ages. The *anin*, or typhoon is a regular occurrence here. It is common custom among Batanes inhabitants to help one another in times of disaster. Existing social institutions like labour co-operatives are naturally ready to help and facilitate concerted community efforts to cope with disasters.

In Ninh Thuan province of Viet Nam, a weather forecasting system needed to warn against incoming droughts did not exist until very recently. Because drought and saline intrusion in groundwater pose a real threat to crops, farmers have developed locally bred short-time seeds and drought- and saline-resistant varieties of rice and maize. To determine in advance whether to sow drought-resistant varieties, local farmers have developed a common indigenous knowledge of weather forecasting based on moon observation and insect behaviour.

In the Mekong River Delta of Viet Nam, local villagers have accommodated traditional farming systems to their changing ecosystem. Because they were facing increasingly severe droughts, villagers have improved their capacity to cope with weather-related disasters and mitigate the impacts of renewed dam building on their environment.

Source: Quang (2017); UNISRD (2008).

Box 4.2. A holistic approach to agricultural risk management

The OECD framework on risk management in agriculture proposes a holistic approach to risk management which distinguishes different layers of risk with the key idea that a differentiated policy response is required for each layer of risk:

- *Normal risk with high frequency* and low damage result from normal variations in production, prices and weather. They can be managed by farmers as part of a normal business strategy, via the diversification of production or the use of production technologies which make yields less variable. A key policy measure to support the management of "normal" risk is extension advice about on-farm practices to strengthen resilience.

- *Marketable risk with intermediate levels of frequency and damage.* Examples are hail damage or some inter-seasonal variations in market prices. These can be handled through market tools such as futures, private insurance or marketing contracts. Governments can also play a role in creating favourable conditions for the development of such market-based risk management tools, by providing information, regulations, training and other facilitation.

- *Catastrophic risk with low frequency and high damage* affect many or all farmers over a wide area; they are beyond farmers' or markets' capacity to cope. Disasters linked to droughts, floods and tropical storms covered by this study often fall into this category of catastrophic risk. As reviewed in Section 4.3, governments have an important role to play with respect to catastrophic risks.

Source: Largely based on OECD (2016a, 2011, 2009).

4.2. Strengthening prevention and mitigation policies to OECD good practices

Following good practices in DRM, policy measures that encourage farmers and their communities to take prevention and mitigation action can reduce their risk exposure to weather-related disasters (Carter, 2008; OECD, 2016c). These measures often require long-term investment into physical and human capital from national authorities and agencies on the one hand, and from households and their communities on the other hand. Therefore, these prevention and mitigation measures require forward planning. Recent milestone weather-related disasters in the four countries studied have inflicted massive damage and losses to the agricultural sector (Table 3.1 in Chapter 3), let alone human casualties, which are still difficult to estimate. This catastrophic impact points to a lack of prevention and preparedness to these recurrent disasters in these countries.

This section reviews five main measures that governments can employ to reduce risk in the agricultural sector. First, (i) water management infrastructure can reduce exposure to extreme weather events. Governments can also provide (ii) information to farmers to encourage risk reduction on the farm. However, information is not always enough to incentivise changes in farmers' behaviour: (iii) regulations, (iv) economic incentives and (v) insurance products should also be designed to reduce the risk from weather-related disasters.

Planning large infrastructure projects needs increased stakeholder consultations and risk analysis

"Water infrastructures projects should involve a transparent and inclusive decision-making process, in which the full set of costs and benefits for different water users and uses (including ecosystems) are clearly recognised using state-of-the-art cost-benefit

analysis. Drawing on lessons from previous failures to estimate the real costs of these projects could be useful in that regard" (OECD, 2016c). The Sendai Framework for Disaster Risk Reduction also encourages signatory countries to undertake risk assessments linked to weather-related disasters for all new infrastructures being constructed (UN, 2015). The international development community has been active in helping ASEAN countries improve their water infrastructure and agricultural sector to become more resilient to weather-related disasters (Alano and Lee, 2016).

Rice is the staple crop of ASEAN countries and has been traditionally grown in flooded paddy fields. Some water-saving rice production systems are now spreading in Asia, for example the Alternate Wetting Drying system promoted by the International Rice Research Institute and the System of rice intensification developed by Cornell University (Cornell University, 2017; IRRI, 2017). Nevertheless, the current water infrastructure system has been erected with the main objective of developing irrigated paddy fields with an emphasis on building dams and irrigation systems. Storing water behind dams has long been seen as a solution to many water problems and large-scale dam infrastructure has therefore been supported by the international development community (World Bank, 2012b). As a result, up-country dams and reservoirs in the countries under study have been erected with a view to store water for irrigation projects while dams and levees in the river plains have been designed to keep paddy water levels stable from seasonal floods. These investments have allowed some countries to irrigate a substantial share of their agricultural land area. The country list below provides more information on the water management infrastructure in the four countries under study. Some countries in the region are planning to increase the surface of land irrigated progressively to reach full irrigation potential. However, this objective might not be coherent with mitigating the impacts of extreme-weather events on the agricultural sector as farmers might be encouraged to grow water-thirsty crops like rice in areas that are not naturally suited for these crops, but where it has been made possible by irrigation development. Moreover, planning water infrastructure to regulate water levels for rice production in irrigated perimeters does not necessarily cope with extreme flooding and drought, as experienced more frequently in the region.

Large man-made water infrastructure projects in ASEAN countries are implemented by the national government in a top-down manner with little consultation of the ultimate beneficiaries. What is more, only recently have infrastructure planning documents in the Philippines and Viet Nam factored in the additional risks from extreme weather events. As such, institutional improvements are still possible to attain OECD good practices.

The decline in mangroves – a natural form of risk prevention that provides relevant ecosystem services to mitigate the impact of floods and typhoons (Beresnev and Broadhead, 2016) –increases the vulnerability of the agricultural sector in several ASEAN countries. Indeed, research has shown that mangroves can contribute to climate regulation, coastal protection, water quality maintenance, and soil stabilisation and erosion control. Mangroves' ecological function of coastal protection benefiting Thailand has been valued at USD 4.3 billion at 2014 prices (UNDP, 2017b). At present, although the area of coastline under mangrove cover has been shrinking in Asia to make way for more remunerative human economic activities such as aquaculture, agriculture, and urban and industrial development (Richards and Friess, 2016), national initiatives are being supported by the international development community to help restore and expand mangrove cover in these coastal areas (Beresnev and Broadhead, 2016). Nevertheless, the interactions between human coastal activities and mangroves suggest governments should carefully evaluate the coherence of their support policies benefiting coastal aquaculture with mangrove protection objectives.

In addition to the general regional findings mentioned above, the following summarises the specific local situation and policy approaches to infrastructure projects for water management in the four countries under study.

- *Myanmar*: Only a small fraction of Myanmar's abundant surface water resources is currently used, but access to water is limiting: less than 20% of crop land is irrigated (IWMI, 2015). Most of the existing irrigation infrastructure is in disrepair and still lacking in areas that need irrigation. Current water infrastructure in Myanmar is often a legacy from the 1990s when engineers calibrated the dams and reservoirs according to meteorological and hydraulic data of the time, and to manage a flood of a magnitude of 1 in 1 000 years. The Irrigation Department updates the calibration of infrastructure that is due for refurbishing to reflect increased risks linked to weather-related disasters. The MAPDRR emphasises preparedness for new constructions after a disaster but it has not much content related to irrigation and water management. As part of the MAPDRR, the government has established a small grants program to foster small-scale infrastructure development by different agencies. After the 2015 floods that had a strong impact on household food security and income, the World Bank released USD 5 million in 2016 – equivalent to 2.3% of the USD 216 million national budget for irrigation for 2016-17 (Htoo Thant, 2016) – to help cover the costs of local-level agricultural infrastructure. However, the Department of Irrigation and Water Utilization Management has not yet been able to spend the funds made available. In the coastal areas, the MAPDRR also supports local communities to regenerate mangrove forest growth to help protect the coastline (Myanmar, 2016; MOALI, 2017; World Bank, 2017; Centre for Economic and Social Development, 2017).

- *Philippines*: The country's irrigation infrastructure is ageing but the government has now realised its vulnerability to weather-related disasters. Expenditures on general services have started to rise sharply at the end of the 2000s with the most important item being the development and maintenance of infrastructure. Within infrastructure public spending, the major share is devoted to investments in irrigation systems (OECD, 2017b). Overall, 57% of the 3 million ha of potentially irrigated area had been developed in 2015. Almost 90% of this irrigated area is used for rice production. In most cases, the engineering design of existing irrigation systems did not take account of changing climatic conditions. Since 2010 the government has aimed to redesign and strengthen the climate resilience of vulnerable irrigation infrastructure. Still, irrigation perimeters are regularly damaged by floods and typhoons (OECD, 2017a).

- *Thailand*: There is a long history of government planning of water management and infrastructure for agriculture throughout the regions of the country. The potential irrigated area accounts for 20% of total agricultural land in Thailand and rice cultivation takes up 40% to 50% of the total irrigated area in the country (USDA, 2017). However, only half of that potential irrigated area is effectively irrigated (MoAC, 2017b). For the country to cope with the 2016-17 drought season, the government has estimated that an additional 17 661 cubic meters of reservoir water was needed, of which 54% will go to agricultural irrigation (MoAC, 2017a).

- *Viet Nam*: A country historically built on the abundance of water is suddenly confronted with a lack of infrastructure to address severe droughts. Since the 1970s, irrigation and flood protection have remained a major focus of the government. As a result, near 50% of the 9.4 million ha irrigation potential in Viet Nam has been developed. Vietnamese engineers calibrate river levees based on the risk level calculated from past flood occurrences: the infrastructure is meant to be able to cope with a once-in-a-hundred-years disaster. According to the Ministry of Finance, Viet Nam currently spends approximately USD 575-660 million every year to invest in agricultural production infrastructure (MARD, 2017a). International development partners often participate in co-financing these large investments and provide their technical expertise on making them more sustainable (AFD, 2014). One study on dyke infrastructure in the Mekong River Delta calculates that the costs of heightening the dykes to reach rice production objectives do not provide benefits in value addition for rice producers or in terms of reduced pesticide use, thus undermining the government's choice of infrastructure development (Tong, 2015). A similar assessment of the risks linked to environmental disasters and climate change is now compulsory before launching new infrastructure investments. A 2014 irrigation restructuring plan encourages provincial and local governments to plan irrigation needs together with agricultural restructuring: applying water-saving irrigation technologies, reorganising agricultural production towards less water-thirsty crops. These innovations have come in handy in the south of the country but were not enough to deal with the 2016 catastrophic drought and inland salinity intrusion (OECD, 2015a; Quang, 2017; IPSARD, 2011; MARD, 2017b).

Farmer awareness remains low due to limited technical assistance on risk management

Advisory services are a second policy tool employed by governments to reduce risk in the agricultural sector. Risk awareness campaigns are indeed a key organisational measure employed by governments to mitigate disaster risk (OECD, 2014b). Even in OECD countries and partner economies, risk communication tools sometimes have failed and there are always pathways to improve the effectiveness of risk communication policies (OECD, 2016d). In particular, technical advice and information campaigns are key to increasing awareness among farmers about their risk exposure and on-farm risk-reduction strategies. Crop diversification and adoption of resistant varieties is one such strategy. By adjusting on-farm practices, farmers can also manage drought risk (for instance, through rainwater harvesting, drip emitters, low-pressure sprinkler systems, concrete lining of irrigation canals) and flood risk (for instance, through hedgerow planting and management and planting on riparian buffer zones or water margins) (European Commission, 2011; OECD, 2016c, 2014b, 2010).[1] Farmers can also improve the efficiency of their water storage to prevent catastrophic impacts of droughts, floods and tropical storms. For example, establishing water corridors linking rivers, paddy rice fields and irrigation ponds creates a natural safety valve when a community is hit by excess or lack of water (OECD, 2010). Allowing groundwater to recharge during times when water needs are low also fosters the resilience of the water system, and the dependent farmers' livelihoods, when catastrophic shocks occur.

Awareness among Southeast Asian farmers of the practices to reduce risk exposure is relatively low, and technical advice provided to farmers on how to strengthen resilience is variable (SEARCA, 2017). A recent UN survey on land degradation, for instance, found that most farmers were not aware that increasing input use was degrading land quality (Economics of Land Degradation Initiative, 2016; EEPSEA, 2017). In several countries, extension services are encouraging crop diversification, a useful risk management strategy. However, other on-farm practices to strengthen resilience – such as the adoption of resistant seed varieties and efficient water use – are less widely promoted. This is due in part to the variable quality of extension advice, which depends on the capacity of the local extension agent, the level of remoteness of the farming community, and the civil society and agribusiness networks that can help transfer knowledge and information. In some cases, it is also driven by a limited focus on research and development on resilience; however, many on-farm practices and tools that increase resilience could be disseminated without further research. A final barrier to awareness may be extension services' top-down approach in certain countries and focus on productivity-enhancing measures, which may misadvise farmers and increase their vulnerability if local conditions are not taken into account.[2]

In addition to the general regional findings mentioned above, the following summarises the specific local situation and policy approaches to raising farmers' awareness on risk management to weather-related disasters and climate change in the four countries under study.

- *Myanmar*: Advice on risk management practices is not commonly shared with farmers in Myanmar. Some resilient practices are promoted in certain areas – for instance, alternate wetting and drying technologies are now employed by some rice farmers. Moreover, lack of policy support for rice production has also led market forces to shift land and labour to pulses, which require less intense use of water than paddy rice. However, there remains significant scope to increase adoption of resilient, low-cost practices at the household level (World Bank, 2017; GiZ ASEAN SAS, 2017). Adoption rates of drought-resistant seed varieties are also low: they are reportedly used on only 120 000 ha. Limited adoption of resilient practices is in part due to capacity constraints of extension services – one extension worker is responsible for a territory covering 2 800 ha on average (GiZ ASEAN SAS, 2017). In an effort to overcome this information gap, MOALI is currently setting up call centres for farmers to receive advice on crop management systems. Several mobile phone apps are also being disseminated (for instance, Greenway advises approximately 28 000 farmers on the best time to plant, weather information, etc.). It is not clear whether risk management will be mainstreamed in such programmes as they scale up.

- *Philippines*: Low-income farmers in the Philippines have limited awareness about their risk exposure and climate change, but plans to strengthen extension's focus on climate related issues are gathering momentum. Agricultural extension has been centralised for many years and mainly focused on increasing production. It was devolved to local government units in 1991 at the provincial, city, and municipal levels. With the advent of climate change and increasing frequency of weather-related disasters, extension concerns now include DRR. Local Government Units however are in general technically ill equipped, have budget and manpower constraints, and have other priorities that affect their capacity to provide advice on resilience issues. Private sector initiatives to incorporate risk awareness in their extension services are seen as a potential way of addressing this gap. Various media (e.g. radio programmes, smart phones, printed materials, technical bulletins) are also increasingly used to disseminate information on new varieties, technologies, and good practices to cope with climate change (OECD, 2017a; SEARCA, 2017).

- *Thailand*: Though resilience is not a central priority of extension services, Thai farmers are provided with information on select practices that reduce risk. According to development partners, Thailand's extension services focus primarily on helping farmers reduce their costs of production in an effort to boost productivity. Such advice could include measures that reduce vulnerability (e.g. promoting efficient water use in drought-prone areas), but it would also include measures that increase productivity in the short-term at the expense of land degradation and higher risk in the long-term (e.g. promoting rubber plantation or maize for its fast economic returns in dry erodible highlands of the North and Northeast). Notwithstanding these potential concerns, extension has provided many farmers with advice on crop diversification, an effective risk management strategy in many instances. As part of the "New Theory" from the Royal Initiative of the Philosophy of Sufficiency Economy, farmers are encouraged to divide their farmland into four parts: (i) 30% for a pond to harvest rainwater and grow fish; (ii) 30% for rice cultivation for household consumption; (iii) 30% for other crops (for consumption and to sell); and (iv) 10% for accommodation, livestock, etc. (Chaipattana Foundation, 2017; UNDP-TH, 2017). While not appropriate for all regions, the principles promoted in this initiative could help smallholder farmers in more fragile agro-ecological settings or lacking market outlets to manage their risk exposure more effectively. Moreover, Thai farmers receive advice on sustainable practices in most villages from volunteer "soil doctors" (Land Development Department, 2017; UNDP-TH, 2017). Lastly, alternate wetting and drying technologies are promoted in some areas.

- *Viet Nam*: Vietnamese farmers receive some information about practices to reduce risk exposure through the loudspeaker system and through extension, but it is not always effective and limited in scope. Viet Nam's loudspeaker system conveys information to citizens at dawn and at dusk in all but the most remote parts of the country. This loudspeaker network is used to convey information on suggested cropping practices, weather conditions, water availability, and on disaster risk management. Radio and television are other means of information sharing with the potential to reach many farmer households (FAO-VN, 2017). However, recent assessments show that the weather forecast and preparedness information prepared for the 2016 catastrophic drought did not reach the farmers affected, although it had been disseminated on various information channels (FAO, 2016a). Farmers also receive some advice on risk management issues – in particular, on crop diversification – from Viet Nam's extension curriculum. However, other issues – such as on-farm measures to manage water resources and the use of resistant seeds – are less well covered (FCRI, 2017; MARD, 2017b; FAO-VN, 2017; OECD, 2015a).

Regulations orient commodity choices with expected indirect impacts on water use

Water allocation rules do not allow an efficient and sustainable use of water by agriculture

Lessons learned from the OECD show that well-enforced, short- and medium-term regulations can help mitigate the onset of water scarcity and overabundance. This section focuses on two regulatory areas that are particularly relevant for water-related risks: water allocation and land zoning. The subsequent section on incentive schemes then looks in more detail at good practices for water pricing.

OECD analysis highlights that regulations to effectively allocate water in the long term – and adjust in the short term – is needed to mitigate the risk of severe droughts. "If the state of water shortage is recognised as critical, then this can trigger a set of water restriction rules, eventually combined with short-term water supply responses, such as supporting a river's flow by releasing increasing volumes of water from a dam or allowing for groundwater pumping. Short-term water restriction rules can concern all water users, but especially those that rely on freshwater withdrawals such as agriculture, urban and industrial users. In certain cases, following or simultaneously to the short-term water restrictions, flexible mechanisms for reallocating water across farmers or a broader set of users can be implemented. In practice, this usually takes the form of water quota trading or water auctions by public agencies (OECD, 2016c). However, robust water allocation arrangements are a necessary foundation to sustainable water management in the agricultural sector. Water allocation arrangements need to be adapted to specific location conditions, but they also have to be robust and adaptive enough to perform under both typical and extreme conditions. The allocation regime should also be able to adjust to changing conditions at least cost over time. OECD (2015b) has developed a "Health Check" tool to review current water allocation arrangements. It enables to assess whether the elements of a well-designed allocation regime are in place and to identify areas for potential improvement.

Water allocation in the four ASEAN countries under study is mainly decided at central level as part of a national planning for water use by the different sectors of the economy. Central governments in Myanmar, Thailand and Viet Nam allocate the overall volumes of water available in the upriver dams for the different fluvial and irrigation basins. The agricultural sector is recognised as a big water user in ASEAN countries, but it is not always ranked as a priority user in periods of drought. Conversely, when flooding occurs, national authorities in ASEAN might have to decide whether to flood agricultural land in order to protect industrial and urban areas (FAORAP, 2017; MARD, 2017b) The general lack of equipment to measure water usage in irrigated perimeters is likely to be a reason to rely on indirect policy instruments to regulate water volume use. The World Bank's Enabling the Business of Agriculture (http://eba.worldbank.org/) water indicators measure key elements within the legal and regulatory frameworks that impact farmers' access to sufficient quantities of water, at an adequate quality level and at the time and location needed for crop production. The integrated water resources management index measures legal mandates including the establishment of basin-level institutions, water planning, the development of information systems, and source protection. The individual water use for irrigation index measures legal requirements for water abstraction and use permits as well as their enforcement. These indices suffer from weaknesses in the method used for data collection and their construction. Nevertheless, they are interesting indicators to benchmark the regulatory environment for water use in the agricultural sector. Whereas the Philippines and Viet Nam compare favourably with selected OECD countries also covered by this ranking, Myanmar is at the bottom of the latest ranking and Thailand is the third worst performer (Table 4.1).

**Table 4.1. World Bank Enabling the Business of Agriculture water indicators
for selected ASEAN and OECD countries (2017)**

Country	Water indicator rank out of 62 countries studied	Integrated water resources management index (0-29)	Individual water use for irrigation index (0-20)
Spain	1	28	18.5
Mexico	2	29	16.5
Korea	9	26	12
Italy	10	20.5	15.5
Greece	12	27.5	10.5
Poland	13	23	13.5
Philippines	17	18	14.5
Netherlands	20	23.5	9.5
Denmark	24	23	8.5
Viet Nam	27	18	11
Chile	28	17.5	10
Turkey	51	14	0
Thailand	60	3.5	0
Myanmar	62	1.5	0

Note: The integrated water resources management index measures legal mandates including the establishment of basin-level institutions, water planning, the development of information systems, and source protection. The individual water use for irrigation index measures legal requirements for water abstraction and use permits as well as their enforcement.
Source: World Bank Enabling the Business of Agriculture data base, http://eba.worldbank.org/.

In addition to the general regional findings mentioned above, the following summarises the specific local situation and policy instruments for water management in the four countries under study.

- *Myanmar:* Myanmar is the worst ranked country studied by the World Bank's Enabling the Business of Agriculture water indicator (Table 4.1). This indicates that regulations for institutional and individual use of water resources in the agricultural sector can be improved greatly.

- *Philippines*: The Philippines is ranked 17th out of 62 countries studied by the World Bank's Enabling the Business of Agriculture water indicator (Table 4.1), above the Netherlands and Denmark.

- *Thailand*: Currently, the Thai Ministry of Agriculture and Co-operatives (MoAC) and its Irrigation Department provide a rigorous allocation of reservoir water during the dry season, thus restricting water discharge in a transparent and consistent way (MoAC, 2017a, 2017b). Thailand has approached FAO to seek assistance on policy development addressing floods, landslides and droughts (FAO, 2016b). Thailand is the third worst ranked country studied by the World Bank's Enabling the Business of Agriculture water indicator (Table 4.1). This indicates there is a large scope to progress in reaching an enabling regulatory environment for water use by agriculture.

- *Viet Nam*: Viet Nam is ranked 27th out of 62 countries studied by the World Bank's Enabling the Business of Agriculture water indicator (Table 4.1), above Chile and Turkey.

ASEAN countries use less-effective land zoning instruments as policy proxies to regulate water use in agriculture

OECD research has also shown that risk-informed land use policy decisions are key to steer human activities away from locations where risks have a high likelihood of occurring (OECD, 2014a). The difficulty of devising relevant risk-informed land use policies lies in the high level of detail on risk information needed to inform single land use choices. Enforcing these policies at a very local level is an additional challenge. For agriculture to become part of the solution to weather-related DRM, risk-informed agricultural land zoning can orient agricultural practices towards production systems that are less water-thirsty and change the hydrological properties of a surface water catchment. Thus, appropriate agricultural land zoning that is enforced has the capacity to provide potential drought or flood mitigation and protection to areas downstream. The restoration of floodplains and wetlands can store water in periods of high or excessive precipitation for use in periods of scarcity in addition to providing other ecosystem services (Largely drawn from OECD, 2016c). In Southeast Asia, a large surface of agricultural land traditionally under rice paddies can similarly provide mitigation benefits during floods: flood alleviation thanks to the retaining capacity of bounded rice fields, groundwater recharge, and soil erosion and landslide control on sloped land (OECD, 2010). If governments have to take the decision to flood agricultural land to protect the economic interests and assets of other sectors of their economy, farmers should be adequately compensated for their losses.

From an agricultural productivity perspective, land zoning and the promotion of specific crops adapted to a given agro-ecological setting can effectively orient farmers towards more environmentally friendly farm practices if the zoning is conducted scientifically and no other market distorting policy support provides a conflicting incentive. Encouraging farmers to grow crops that suit a particular agro-climatic zone is consistent with the economic theory of comparative advantage and can contribute to reducing the stress to agricultural resources such as water and soil. Technical advice to farmers on optimal agricultural practices for their agro-ecological zone can spread water-saving production techniques. Conversely, encouraging farmers to grow crops that are not suitable to a particular agro-climatic zone can be harmful for the environment, putting additional stress on land and water.

The ASEAN countries studied rely mainly on land zoning and the promotion of specific crops adapted to a given agro-ecological setting to regulate water volume use by farmers rather than using robust and adaptive water allocation regimes, or restricting water extraction or irrigation use. The principal objective of land zoning policies in some of the four countries studied is to encourage farmers located in the most fertile areas to produce the crops that will bring the highest yield or market return in these locations, while suggesting more resistant crops in areas regularly suffering from drought. Thus, the objective of land zoning is predominantly to increase or maintain the country's agricultural production rather than encouraging risk management by farmers. In fact, not only is the zoning not based on past weather conditions but it rarely accounts for climate change and weather-related risks, although some progress on taking account of these risks has been made, as detailed in the country-specific examples below. Yet, this indirect mode of adapting crops to ecological setting is a less effective way of orienting farmers towards water savings than directly restricting water use in the region's water stress hotspots. Several of the countries studied consequently fall short of OECD best policy practice of robust and adaptive water allocation regimes, or restricting water extraction or irrigation use as more direct and effective policy instruments to achieve the same objective. Land zoning can help identify the areas most prone to being hit by catastrophic drought, flooding and storms. In practice, it is not always compulsory for farmers in the countries studied to follow the crop production system recommended by the land zoning regulations in a given agro-ecological area. Farmers are still free to obey market signals to choose which crops to grow (though financial penalties are imposed in certain instances, as discussed in the next section).

In addition to the general regional findings mentioned above, the following summarises the specific local situation and policy approaches to zoning regulations in the four countries under study.

- *Myanmar*: The country is completing its mapping of disaster-prone areas but has no plan to orient agriculture production according to the diversity of its agro-ecological zones. As part of the implementation of the MAPDRR, the government has completed the flood risk maps in Kathat, Kalewa, Mandalay and Hpa-an townships (Myanmar, 2016). Myanmar will implement marine and coastal spatial planning for specific coastal areas under the lead of the Ministry of Natural Resources and Environmental Conservation and with the collaboration of MOALI, among other collaborating national and international agencies (FAO, 2016b). On the other hand, the country does not have any land zoning plan to suggest the most appropriate farming practices to farmers according to their agro-ecological setting (MOALI, 2017).

- *Philippines*: The country's land management regulations are not yet geared to factor DRM. The Philippines' land tenure reform since 1988 has emphasised the transfer of land to formerly landless households. Only in 2014 has the National Land Use Act (NLUA) proposed to harmonise conflicting land laws and to regulate spatial planning, in particular, with a view to disaster preparedness and prevention. As of December 2016 the NLUA had still not been signed into act because of the highly politicised nature of land tenure issues (OECD, 2017a).

- *Thailand*: The country is starting to encourage rice farmers to diversify their production to decrease its rice overproduction. Because the country produces much more rice than it can consume, the government has set up a Committee on Integrated Rice Farming Practices to suggest the areas in the country in which to limit rice production. This agricultural zoning would incentivise rice farmers to change cropping system according to their agro-ecological setting. These recommendations are not mandatory and do not lead to penalties if not followed (MoAC, 2017b). The Thai Ministry of Agriculture and Cooperatives has set up an online geographical information system called AgriMap providing recommendations on crop choices at national level (hwww.moac.go.th/agri-map/). This tool constitutes the basis for the agricultural zoning policy of the government.

- *Viet Nam:* The country's agricultural zoning plan is key to reaching the economic and social objectives of the national five-year plans. Viet Nam has an elaborate land zoning plan. Because it is the country's staple crop and employs a large share of rural labour in its production, distribution and processing, policies supporting rice production are perceived as contributing to food security as well as social stability objectives. Overall, 3.8 million ha of land are allocated for the sole production of rice, which leads to specialised infrastructure and the promotion of productivity enhancing practices. Rice farmers located on land assigned for rice production can request an official authorisation to change cropping system but these are rarely given to attain the national total rice surface objective (OECD, 2015a). IPSARD has conducted a foresight study suggesting that the country could reduce the area under rice to only 3 million ha and still be self-sufficient in rice but the government and Party have decided to keep the previous objective in place and secure social stability (IPSARD, 2017, 2011). In 2016 the government has introduced a specific decree requiring farmers wishing to convert land assigned to rice out of rice production to pay a conversion fee to the local government. In response to the 2016 drought in the South of the country, the government has issued regulation encouraging farmers in the drier areas of the Mekong River Delta, Central Highlands and Northern Highlands to switch from irrigated rice to rain-fed maize cultivation with a subsidy of VND 3 million (USD 132 million) per ha to purchase maize seed (MARD, 2017a).

Exceptions notwithstanding, many economic incentives increase vulnerability

Depending on their design, economic incentives such as agricultural support measures can discourage or encourage resilience. "Subsidies and commodity price supports [for instance] can affect farmers' production decisions and, in certain circumstances, increase their exposure to drought and flood risk by encouraging them to cultivate high-risk land, high-risk crops or divert them from adopting a more diverse range of activities" (OECD, 2016c). Subsidised water (Box 4.3) is another support measure that – along with weak water allocation regulations – can discourage efficient water use and increase vulnerability to droughts. In some contexts, however, certain economic incentives may actually encourage resilience. For instance, when the costs of investment are too high, targeted, short-term agricultural support can encourage farmers to invest in resilient technologies.

Similar to OECD countries, ASEAN governments use a mix of economic incentives; many increase farmers' vulnerability to weather-related disasters, though some encourage resilience at a smaller scale. In several countries, market price support measures – such as import restrictions and minimum prices – encourage farmers to engage in activities that are not suited to local conditions. Concessional loan terms and subsidised insurance (as discussed in the next section) for rice farmers are also common. However, some economic incentives are also encouraging resilient investments. For instance, in some countries, concessional credit is offered to farmers that invest in efficient irrigation schemes. The challenge of course is that, in the absence of oversight, encouraging an efficient irrigation system can also lead to increase water consumption via changes in crops or extension of land.

In addition to the general regional findings mentioned above, the following summarises the specific local situation and government agricultural incentives that might be detrimental to reach long-term sustainability objectives in the four countries under study.

- *Myanmar*: Several incentives to encourage rice production may increase vulnerability in Myanmar. For instance, rice production is currently encouraged – regardless of local conditions – by concessional loan schemes from the main institutions for agricultural credit such as the Co-operatives Department and the Myanmar Agricultural Development Bank (MADB). The Co-operatives Department offers loans to co-operative members[3] for up to 10 acres, 3 acres of which must be for paddy. The MADB provides loans of up to MMK 150 000 (USD 111) to rice farmers, versus only MMK 20 000 (USD 15) if farmers grow other crops. Some farmers take the rice loan and then use it to plant another crop, i.e. enforcement is weak; however, such schemes may still nudge other farmers towards rice production, even in areas that are not well suited to it (MOALI, 2017). A second potential incentive for rice production that may emerge in the future is a minimum price. "In 2013, Myanmar adopted the Farmers' Rights Protection Act that sets out the possibility for the introduction of minimum prices for agricultural commodities such as rice. However, the World Bank reports that implementation details are not yet clear, and, given limited fiscal resources, a public procurement system such as that used in some other countries is unlikely to be feasible" (World Bank, 2014, as quoted by OECD, 2017c).

- *Philippines*: Several incentive schemes in the Philippines may distort farmer incentives and weaken their resilience in vulnerable areas. Of particular concern are market price support measures in the form of import quotas for rice and high tariffs for sugar and animal products (OECD, 2017a). For instance, "rice imports are controlled by state-owned enterprises – or regulated monopolies – that control the quantity of imports […]. These policies limit the supply of rice to the market and provide a means for governments to increase producer prices" (OECD, 2017a).

- *Thailand*: While some agricultural support measures are distorting farmer incentives, others are encouraging resilient investments. The Thai government recently abandoned a minimum price scheme for rice that may have distorted farmer incentives to grow more resilient crops (OECD, 2017c). However, to address falling rice prices and strong lobbying from farmers' interest groups, the government reinstated target intervention prices set below the market prices for different local varieties of rice and offered new cash incentives and subsidised loans to rice farmers to stockpile rice in late 2016 (Peel, 2016). Thailand is also encouraging some resilient practices through economic incentives. For instance, to promote diversification and water access, the government pays two-thirds of the cost of building a new farm pond (RID, 2017). Moreover, to implement the country's zoning plan, farmers are offered economic incentives to reduce rice production and switch to another crop suitable for the local conditions (MoAC, 2017b).

Box 4.3. Water pricing

Water pricing is an important economic instrument for encouraging efficient water usage and reducing the risk of drought; however, identifying the appropriate price can be difficult and should form part of a broader water allocation and management system. "Administrative water pricing is typically implemented by a government or collective agency that provides a water service and attempts to recover at least part of the cost. Reducing water demand […] through a price increase [is challenging, as it] requires some knowledge of the price elasticity of water demand, which can vary a great deal across time. In addition, water price increases may have negative equity implications for farmers, leading to regressive redistribution through water demand. Finally, political pressures from the agriculture sector are susceptible of undermining the development of such an instrument. Broader institutional aspects can also play a role" (OECD, 2016c). Such challenges notwithstanding, the OECD recommends that agricultural water charges cover the full supply costs at a minimum. Ideally, charges should reflect the opportunity cost of withdrawals and be accompanied by social and adjustment policies to compensate low-income farmers (OECD, 2016b).

The price of water is very low – or sometimes free – in Southeast Asian countries, encouraging overuse and increasing the region's vulnerability to drought. In countries that charge water use, the fee is normally based on surface area – as opposed to units consumed – which is encouraging overuse.

- *Myanmar:* The price of water for agricultural producers in Myanmar is very low and charged according to land area. Prices charged in government irrigation systems range from MMK 2 250 to 22 500 (USD 1.7-17) per ha; the actual cost of providing water is much higher, as much as MMK 250 000 (USD 185) per ha in pumped irrigation systems (IWMI, 2015).

- *Philippines*: Farmers were previously charged an irrigation service fee to cover operational and maintenance costs of water irrigation systems, but this fee has been abolished in 2017. A fund worth USD 45.9 million has been constituted to cover the national irrigation costs in 2017 (OECD, 2017b).

- *Thailand*: Water is provided free of charge to Thai farmers by the Royal Irrigation Department (RID, 2017), exacerbating the country's water overuse problem. The government is starting to encourage farmers to reduce their input costs and water usage; this may help reach the objectives of strengthening agricultural resilience and increasing farmers' productivity.

- *Viet Nam*: Water overuse – a growing concern in Viet Nam – is exacerbated by the low cost of water. High subsidies – in particular, for rice farmers – discourage efficient use of water and increase the sector's vulnerability to drought. Viet Nam is currently developing a new law on water irrigation, which will likely vary according to crops grown and surface area (MARD, 2017a; IPSARD, 2017). "Re-establishing a water fee based on a per-unit-of-water charge rather than a per-hectare charge, as previously applied, would encourage greater water use efficiency (OECD, 2016a).

- *Viet Nam*: While market price support measures may increase vulnerability among Vietnamese farmers, other economic incentives and penalties appear aligned with resilience objectives. In terms of market price support, "producers of import competing commodities such as beef and veal and sugar cane are protected by tariffs. [Moreover,] farmgate rice prices are supported by a subsidy to rice purchasing enterprises for the temporary storage of rice during harvest and establishment of target prices which vary between regions and crop season with the objective of providing farmers with a profit of 30%" (OECD, 2016a). These measures may encourage farmers to produce such products, even in vulnerable regions. On the other hand, the country's rice and coffee zoning programme reportedly takes vulnerability into account and is enforced with financial penalties (MARD, 2017b). If farmers decide to grow something other than rice or coffee in the designated zones, they are not eligible for financial support when a disaster occurs. Moreover, economic incentives are offered to encourage efficient water usage in some areas: concessional credit schemes are offered to coffee farmers who invest in water-saving irrigation systems (MARD, 2017b).

Agricultural insurance is spreading in some countries, but design issues constrain its effectiveness

When designed effectively and co-ordinated with other financial tools and a broader risk management strategy, agricultural insurance may serve two key roles: reducing risk exposure and transferring catastrophic risk to insurance markets in a cost-effective way (Box 4.2).

A first prerequisite to reducing risk exposure is an effective information campaign. By educating farmers about how insurance works and how useful it can be to hedge their risk, awareness about risk exposure will increase; in turn, better information to farmers on available agricultural insurance products may encourage them to reduce their risk exposure in certain contexts – for instance, to minimise their losses or qualify for a lower premium.[4] While the impact of insurance on reducing risk exposure has been limited in practice, the literature suggests potential for risk reduction spillovers in the context of information about risk exposure. Affordable opportunities to reduce risk – for instance, by bundling insurance with drought-resistant seeds – may also be appropriate in certain contexts (Carter et al., 2008; Ward et al., 2015; Surminski, 2013).

The cost of insurance may also affect insurance penetration rates and incentives to reduce risk. For instance, if insurance is heavily subsidised, it can distort farmers' incentives to reduce their risk exposure (OECD, 2016c). On the other hand, if the insurance premium is above farmers' willingness to pay, farmers will not purchase insurance at all. The challenge is thus to strike a balance that maximises the risk transferred while minimising the level of risk that farmers are induced to take. Governments have a key role to play in this regard; a clear framework for the allocation of risk and responsibilities and support in the development of insurance markets is often needed (OECD, 2016c). Moreover, as highlighted in the OECD Recommendation on Disaster Risk Financing Strategies (OECD, 2017d), governments should also ensure insurance is complemented by effective land-use regulations as well as targeted investments in preventive measures.

To reduce risk exposure and cost effectively manage the remaining risk, careful design of the insurance policy is also critical. Historically, most agricultural insurance products were indemnity-based, i.e. pay-outs were based on observed losses at the plot level.[5] However, many developing countries are moving towards index-based products – which link pay-outs to indices such as rainfall levels or area yields recorded at a more aggregated level. The advantage of index-based policies is that they are less expensive to implement (due to lower monitoring costs and less moral hazard), particularly in countries with predominantly small farms. An important drawback, however, is that index-based policies only insure a proxy of farmers' actual losses: this means that the pay-out farmers receive may not be sufficient for losses incurred or may even be too generous (OECD, 2016c). Many countries have found that a hybrid product that combines traditional crop insurance with an index-based insurance product offers particular advantages for resilience: it can (i) reduce transaction costs by standardising the distribution of pay-outs; (ii) ease liquidity-related barriers (by providing partial cash pay-outs quickly based on losses recorded in satellite images); and (iii) curb moral hazard by linking pay-outs to shocks at a more aggregate level (World Bank, 2014a; 2013; 2012a). However, notwithstanding efforts to lower costs, most insurance products are still unaffordable for low-income farmers. Subsidy schemes can help to increase take-up, but should be carefully designed to ensure that the incentives remain for farmers to reduce their risk. Alternatively, insurance can be focused on wealthier farmers and a disaster-linked cash transfer programme can be developed for poorer farmers. Lastly, the success of insurance products also depends on their level of interaction, substitutability and complementarity with other financial products such as savings, loans and cash transfers, which are discussed further in other sections of this report.

Agricultural insurance markets in Southeast Asia are developing, though current policies limit resilience due to design flaws and a focus on water-thirsty crops. A review of agricultural insurance in ASEAN countries identified that markets remained underdeveloped in most countries and achieved low rates of penetration (GFDRR, 2012). Among the case countries for this study, the Philippines and Thailand are more advanced and both offer indemnity-based policies for selected agricultural products. Though both schemes offer advantages, they also have design constraints that are limiting their potential benefits in terms of risk reduction and risk transfer. Myanmar and Viet Nam have yet to develop agricultural insurance products at the national scale, though several pilots have been initiated.

A common challenge in the development of agricultural insurance markets for many Southeast Asian countries is the high-level – and often high correlation across regions – of risk exposure. In particular, disasters occur regularly in certain locations, impeding the pooling of risk in the form of private insurance. Public insurance policies therefore have an important role to play in covering such risks. A regional risk pooling vehicle has also been proposed as a potential tool for managing such risks.[6] Such a pool would provide public funds needed to secure the trust of private reinsurers who could help establish a sound financial basis for a private agricultural insurance market in the region. The ASEAN Cross-Sectoral Coordination Committee on Disaster Risk Financing has recently started negotiating the creation of such a regional risk pooling mechanism (Artemis, 2017).

In addition to the general regional findings mentioned above, the following summarises the specific local situation and policy approaches to agricultural insurance in the four countries under study.

- *Myanmar*: While several small pilots are ongoing (Swiss Capacity Building Facility, 2015), the agricultural insurance market in Myanmar is largely undeveloped. There is interest in developing an insurance scheme among government counterparts, but it is not a policy priority for the immediate future. To design an index-based policy, Myanmar would first need to strengthen the country's information system (Kloeppinger-Todd and Sandar, 2013; MOALI, 2017).

- *Philippines*: The Philippines Crop Insurance Corporation offers indemnity insurance for damage from natural calamities and pests and diseases to farmers producing rice, maize, high-value commercial crops and livestock, among others. As with most indemnity-based policies, the pay-outs are subject to insurance fraud and lengthy waiting periods (normally 60 days) that impede efficient – and thus often resilient – spending on farm expenditures. Moreover, while most farms are required to cover the full insurance premium, several programmes also provide subsidised insurance – in particular, for rice and corn production – to registered subsistence farmers and certain loan recipients. Though partial subsidies can help increase take-up levels in certain contexts, large preferential subsidies such as these can encourage farmers to plant specific crops, even if they are not well suited to local climate conditions. In collaboration with several development partners, the Philippines has started to pilot area-based yield index insurance and weather index-based insurance products, which would help overcome the limitations of indemnity insurance. An additional challenge that future efforts will need to tackle is that coverage levels remain low – only 7% of total agricultural land is currently insured (DA, 2015, as quoted by OECD, 2017a).

- *Thailand*: The National Rice Insurance Scheme – a new indemnity-based insurance scheme for damage from natural disasters[7] for rice farmers was launched in 2014 and has expanded significantly in the last couple of years, from 240 000 ha in 2015 to 4.8 million ha in 2016 (nearly half of Thailand's 10.1 million ha of farmland). This dramatic increase was achieved by shifting from selling insurance with a 60% subsidy directly to farmers to fully subsidising the insurance and incorporating it into loans from the Bank for Agriculture and Agricultural Cooperatives[8] (Oxford Business Group, 2016; Chantanusornsiri, 2016; Rattanayod, 2016). While such a system overcomes low willingness to pay for insurance (Nabangchang et al., 2014; UNESCAP, 2015) and has increased the number of farmers that are covered, it may be less effective at raising awareness and encouraging farmers to reduce their risk exposure as farmers are less active in the transaction and thus risk information is not necessarily transferred. Moreover, the scheme does not accurately reflect shocks suffered at the farm level as a claim is dependent on the government declaring a disaster for the entire area. Thailand has also piloted an area yield index insurance policy that would help to minimise this issue, but it is still under development (Chantanusornsiri, 2016; Rattanayod, 2016). Looking ahead, the scheme should also be expanded to cover other agricultural products, so as not to distort incentives and encourage rice production in vulnerable regions.

- *Viet Nam*: Viet Nam's agricultural insurance market is still in early stages. A pilot insurance scheme was launched in 2011 to insure paddy (through an index-based policy) and livestock and aquaculture (through an indemnity-based policy) with varying subsidies according to income levels. However, the scheme was not extended due to both demand-side constraints (low take-up levels as a result of low coverage and limited awareness) and supply-side constraints (high expense for the government and limited buy-in from the private sector). MARD is currently evaluating plans for insurance development in the future (FAO, 2016a; Lotsch et al., 2010; MARD, 2017b; Nguyen, 2017; OECD, 2016c; Tinh, 2017).

4.3. Institutional and household preparation activities are picking up

This section reviews two key policy instruments to help farmers and their communities prepare for the risks that cannot be eliminated by prevention and mitigation measures: (i) weather and hydrological information systems and (ii) and precautionary savings.

Weather and hydrological data gathering is increasingly accurate but not always widely disseminated

A robust weather and hydrological information system allows governments to identify and map the risks from weather-related disasters. A performing hydrometeorological information system is therefore and important part of the DRM cycle (OECD, 2014b). In the agricultural sector, timely weather and water level information allows farmers to plan evasive action if forewarned of an incoming drought, flood or storm (OECD, 2016c). It also helps insurance providers make more accurate calculations of the risk exposure of their customers, and thus calibrate their insurance products. A dense network of hydrometeorological stations is needed to collect information on weather and water indicators. These indicators allow meteorological services to calculate the risk of weather-related disasters from past occurrences and to forecast the onslaught of catastrophic droughts, floods and tropical storms. Innovations based on satellite imagery could also improve weather forecasts. Related to weather and water information systems, improved statistical data on land use and crop production are also important elements of a robust agricultural information system and can support the design of robust, efficient insurance schemes. However, just as important as the data needed for weather forecasts is the information network to disseminate the forecasts to the rural areas and farmers. The section above on information campaigns to build the awareness of farmers to risk management has already dealt with information dissemination channels to farmers. The same information and communication technologies can be used to reach farmers for long-term awareness raising about risk management and short-term early warning systems.

ASEAN countries have become aware of the importance of a robust weather and hydrological data collection system. In ASEAN, major catastrophic events like the 2004 Indian Ocean Tsunami, the 2008 Nargis cyclone in Myanmar, the 2011 floods in the Central valley of Thailand and the 2013 typhoon Haiyan in the Philippines have raised governments' awareness of the importance of gathering weather and river level data. Countries in ASEAN are thus improving the coverage of their weather and water information systems but conveying relevant and timely hydrometeorological information to the farmers seems to be a general Achilles heel. One product of this new paradigm was the Regional Integrated Multi-Hazard Early Warning System for Africa and Asia (RIMES), based in Bangkok, Thailand (www.rimes.int/). Its mandate is to gather and disseminate data to forecast weather-related disasters, among other extreme events. ASEAN countries also have their own ASEAN Climate resilient network knowledge forum to gather and share information on how agriculture can be better prepared to weather-related disasters (http://asean-crn.org/). The ASEAN Food Security Information System set up by ASEAN countries, China, Japan and Korea includes an early warning system on any foreseen difficulties in production of major food crops that will affect food security (www.resakss-asia.org/partners/afsis/). Finally, ASEAN's Coordination Centre for Humanitarian Assistance is in charge of sharing good practices and co-ordinating regional responses to disasters, a data information centre, and capacity building of national staff on disaster management (FAORAP, 2017). In parallel, the four countries studied have developed appropriate organisations and regulatory frameworks on the topic of water and weather forecasting services (Table 4.2). In addition to public weather and hydrological information services, the private sector and local communities can also participate in sharing relevant information to farmers (FAO, 2016b). For example, allowing upland and lowland communities to share information on their weather conditions and farming practices would enable upland communities to interpret upland rainfall intensity, thus providing real-time warnings to the lowland villages of possible flooding conditions.[9]

In addition to the general regional findings mentioned above, the following summarises the specific local situation and policy approaches to improving the hydrometeorological information system in the four countries under study.

- *Myanmar*: The country has put a commendable emphasis on improving its network to collect weather and water data. This network has become fairly effective; though it could be densified further (Aye, 2014). The Department of Meteorology and Hydrology (DMH) generates weather forecast and early warnings for cyclones, storm surges and floods. A colour-coded cyclone warning message was started in 2009 to disseminate early warning messages more quickly. As part of the MAPDRR, the Myanmar government is improving its meteorological observation and forecasting with automatic weather observation stations set up in 30 townships throughout the country (Myanmar, 2016). The current network of automatic weather stations allows coverage of a 50 km radius per station. To reach international standards, the network would need to be densified to cover a 5 km radius per station. For disseminating this information, the country is continuing the establishment of a multi-hazard end-to-end early warning dissemination system in selected villages in the Ayerwaddy, Sittwe and Rakhine regions. The media and local communities also disseminate this information to farmers. The DMH has organised the Monsoon Forum as a mechanism to foster information sharing between forecast producers and users. International and bilateral development partners are particularly active in helping the Myanmar government improve its weather and water level information system.

- *Philippines*: The national weather monitoring system has improved but the effectiveness of the early warning system can be upgraded to reach more farmers. Recent catastrophic weather events have highlighted the importance of strengthening the Philippine Atmospheric, Geophysical and Astronomical Services Administration (PAGASA) (UNISDR, 2013a). However, "the effectiveness of early warning systems remains relatively low. Around 70% of farmers in the Philippines receive warnings on tropical cyclones 24 hours prior to the event. But only 10% receive warning on flooding; 12% on continuous rain; 4% on drought; and 13% on temperature increase. The number of weather stations also remains limited." (OECD, 2017a) Filipino local community-based information relays are very common and relatively efficient (Seng, 2016; UNISDR, 2008).

- *Thailand*: The Department of Meteorological and Department of Mineral Resources are in charge of gathering weather and water level information. Once transmitted to the central DRR office at the Department of Disaster Preparedness and Management, the information is transferred to the provinces likely to be affected for local dissemination (Chariyaphan, 2012).

- *Viet Nam:* The country has a long-standing history of gathering meteorological and hydrological data but the targeting of the information disseminated can still be improved (UNISDR, 2013b). The National Hydro-Meteorological Service provides immediate, short-, medium- and long-term weather forecasts. Its information is accurate but more effort can still be made in tailoring the information to the different users, including the various levels of education of the farming population in Viet Nam. Viet Nam is involved in an FAO project on Watching Agricultural Drought Worldwide from Space using the FAO-Agriculture Stress Index System. A pilot project is exploring the possibility of applying FAO's ASIS tools to the Central Coast and Central Highlands of Viet Nam. The project is meant to assess the current operational systems for agricultural drought monitoring and improve it with geospatial information to complement the existing agricultural drought monitoring and early-warning system. The project will start in Ninh Tuan Province and will produce a technical proposal to upscale the pilot system to national level (www.fao.org/climatechange/asis/en).

Table 4.2. Summary of legal and institutional framework on hydrometeorological services in selected ASEAN countries

	Myanmar	Philippines	Thailand	Viet Nam
National Meteorological or Hydrometeorological Service	Department of Meteorology and Hydrology, Ministry of Transport	Philippine Atmospheric Geophysical and Astronomical Services Administration, Department of Science and Technology	Thai Meteorological Department, Ministry of Digital Economy and Society	National Hydro-Meteorological Service of Viet Nam, Ministry of Natural Resources and Environment
Law, decree or other legislative act on meteorology (or hydrometeorology, or similar)	None	Presidential Decree No. 78: Establishing the Philippine Atmospheric Geophysical and Astronomical Services Administration	None	Prime Ministerial Decree
Current level of service provision for different types of services (country self-assessment)	Public weather services: Climate services Hydrological services Tsunami services (partly satisfactory)	Public weather services: Warning services Climate services Aviation services Marine services Agrometeorological services Hydrological services Tailored services to specific economic sectors (no self-assessment)	Public weather services: Warning services Climate services Aviation services Marine services Agrometeorological services Hydrological services Tsunami service (satisfactory: meeting most of the requirements)	Public weather services: Warning services Climate services Marine services Agrometeorological services Hydrological services Air and Water quality (no self-assessment)

Source: World Meteorological Organization Country Profile Database https://www.wmo.int/cpdb/.

Informal precautionary savings are a common risk management tool among wealthier farmers

Precautionary savings are a useful financial tool for farmers to prepare for weather-related disasters. "Defined as the extra savings made by an economic agent due to the presence of a given risk in the future," [precautionary savings can take the form of individual or mutual systems. An important advantage of precautionary savings is that they are flexible and can be used to] manage all risks whatever their origins: weather, markets, accidents, etc. Of course, there are challenges with this approach: [individual risks assessments are subject to uncertainty] and building precautionary savings can be difficult or impossible for those who are in financial stress" (OECD, 2016c). Saving with formal financial institutions is one way to encourage better money management in the context of such challenges.

Precautionary savings is a common risk management tool among wealthier farmers in Southeast Asia, though formal savings accounts are still not widely used. Given low and volatile income levels in many ASEAN countries, the majority of smallholder farmers in the region do not have access to precautionary savings. However, among farmers with higher incomes, savings are commonly set aside to prepare for emergencies such as weather-related disasters. Most keep savings at home, but use of formal savings accounts is increasing. Given the complementarity of various financial instruments to hedge against risk, the prevalence of precautionary savings among wealthier farmers in the region could potentially reduce demand for insurance.

In addition to the general regional findings mentioned above, the following summarises the specific local situation and policy approaches to encourage farmers to use precautionary savings in the four countries under study.

- *Myanmar*: Savings levels in Myanmar are very low, particularly in rural areas. High poverty levels – and thus limited disposable income – as well as high transaction costs are two key barriers in this regards (Kloeppinger-Todd and Sandar, 2013). In an effort to increase savings rates in rural areas, the MADB now requires its loan recipients to open a savings account with MMK 3 000 (USD 2.2). However, less than 10% of farmers in Myanmar have loans with MADB (MOALI, 2017). Scaling up such mechanisms would help increase the low level of savings in rural areas.

- *Philippines*: Precautionary savings are a common approach to manage risk in the Philippines, but formal savings tools are not widely adopted. According to a 2015 survey, the most common savings priority of Filipinos is emergencies: nearly two-thirds (64%) of Filipinos that save have emergency funds. However, access to formal savings opportunities is limited. Among rural Filipinos who save money, nearly three quarters (72%) keep their savings at home. Only one third have savings in a bank account (26%) or with a co-operative (9%). Formal savings rates in urban areas are higher (40% in bank accounts), highlighting challenges in the accessibility of banks in rural areas (Bangko Sentral ng Pilipinas, 2015).

- *Thailand*: Undersaving is a common problem in rural Thailand. According to Patmasiriwat and Hengpatana (2016), 29% of rural households are in debt and 47% do not save enough to cope with future risks. The Bank for Agriculture and Agricultural Cooperatives has made efforts to increase rural savings rates in recent decades (BAAC, 2010). As a result, close to half of Thai farmers now have a formal savings account to qualify for more elaborate financial products like loans and insurance (Oxford Business Group, 2016).

- *Viet Nam*: Most farmers have access to formal savings opportunities in Viet Nam, but such services are still not widely used and savings rates in rural areas remain low. Exceptions in very remote areas notwithstanding, most rural areas are well covered by the Viet Nam Bank for Social Policies, the Viet Nam Bank for

Agriculture and Rural Development and co-operative banks. Nonetheless, savings levels remain low and formal services are not widely used, particularly among low-income rural households (IPSARD, 2017; OECD, 2015a).

4.4. ASEAN governments have honed their capacity to respond to weather-related disasters

This section explores two policy tools needed to respond quickly to weather-related disasters in the agricultural sector, namely: (i) crisis management procedures and (ii) disaster linked cash transfer programmes.

Crisis management procedures are well established in the legal framework for DRM

As described in Section 4.1, the policy guidelines from international organisations recommend setting up legal frameworks and institutions to implement effective DRM. This includes clear protocols on how to manage crises and the responsibilities of different stakeholders when critical decisions are needed quickly under difficult and complex conditions (ASEAN, 2016; Carter, 2008; CFE-DMHA, 2015; IFRC and UNDP, 2014; OECD, 2015d, 2014b; UN, 2015).

All case countries studied have followed best international and regional practices by establishing legal frameworks and institutions to elaborate and implement crisis management procedures. However, the key challenge lies in the implementation of these DRM frameworks and strategies when disaster strikes. The national regulations, plans and activities are then adapted to fit local contexts under the co-ordination of local governments. With the high frequency of weather-related disasters in ASEAN, the frameworks and plans for DRM are unfortunately put into practice relatively often. On the bright side, this allows the government authorities and civil society to learn from experience and improve their crisis management procedures. The Ministries in charge of agriculture and rural development in the region often also have responsibility over hydraulic infrastructure. Thus, the interests of the agricultural and food sectors are taken into account in crisis management procedures although the priority for disaster response is more often saving lives and restoring transport access to devastated areas. Still, better governance and co-ordination between local units of the line ministries is needed to allow this devolution process to bear fruit for rapid disaster response.

When a weather-related disaster occurs in an ASEAN country, the relevant national committee in charge of disasters co-ordinates the activities of the line ministries to organise renewed access to the affected area; emergency evacuation of casualties; emergency distribution of essential food, drinking water, blankets, clothes and medicine; emergency reconstruction of damaged infrastructure. Viet Nam in particular has set up a national reserve with stocks worth USD 3.52 million in 2016 for these essential goods, to be used in times of disasters and epidemics. The national reserve is buttressed on a network of local reserves (MARD, 2017b).

ASEAN countries facing a food security emergency can request the assistance of the ASEAN Plus Three Emergency Rice Reserve (APTERR) set up by ASEAN, China, Japan and Korea. Although existing since the 1970s, APTERR is still struggling to live up to its goal of improving the fluidity and rapidity of emergency rice distribution in its member countries. Its main limitations in terms of emergency distribution are linked to the sometimes slow pace of emergency reconstruction of damaged infrastructure, which remains a national prerogative (Trethewie, 2013).

More controversially, the Philippines has intervened in food markets following a weather-related disaster. The Filipino government set a temporary price freeze in modern and traditional retail markets for basic commodities after the country was hit twice by typhoons Ondoy and Pepeng in 2009. These price caps were meant to limit food price inflation, which hit basic food commodities and vegetables as major producing areas of the country had been hit by the disasters (World Bank, 2011). Rather than setting price caps on local markets, it might have been more efficient to import the goods necessary to replace the production damaged, allowing poor households to ensure their food security.

Disaster-linked cash transfers offer a useful safety net in many countries, though programme design could be improved

Disaster-linked cash transfer programmes are a key financial tool for managing the immediate financial impact of weather-related disasters for low-income households in the agricultural sector. Due to financial constraints and lower levels of financial education, take-up of agricultural insurance products tends to be very limited among low-income households. Disaster-linked cash transfers help to fill this gap in financial coverage and reduce the risk that such households fall further behind when a disaster occurs. By registering participants in advance and establishing a quick cash disbursement mechanism (such as via mobile banking), participants can receive a small cash transfer within a few weeks of suffering losses from a disaster. Such payments enable farmers to purchase new inputs, thus strengthening their resilience by reducing disaster-induced disruptions to production and avoiding the sale of valuable assets. When data on agricultural losses is collected via satellite data and farmers are paid directly through modern technologies such as mobile phones, disaster-linked cash transfers can also be cost effective at scale (World Bank, 2014a, 2013; OECD, 2017a).

Disaster-linked cash transfer programmes are already employed in some Southeast Asian countries, providing an important source of financial relief. However, many programmes are slow to make disbursements and costly for the government. In some countries, loss assessments are undertaken for each farm, which slows the disbursement process and inflates programme costs. Some countries target poor households. However, in others, all farmers are eligible, regardless of farm size and income levels. This is problematic as it may actually increase risk-taking among wealthier farmers that have other forms of insurance. The lack of targeting also increases costs for governments, which are likely to persist in the long term: offering disaster-linked transfers to all farms fosters a culture of dependence, reducing larger farms' demand for insurance and increasing their expectation that the government will support them when a disaster occurs.

In addition to the general regional findings mentioned above, the following summarises the specific local situation and policy approaches to emergency cash transfers after a disaster in the four countries under study.

- *Myanmar*: Myanmar does not have a disaster-linked cash transfer programme, but existing frameworks may support the development of such a scheme in the coming years. As a first step towards developing a disaster-linked cash transfer programme, the Ministry of Social Welfare, Relief and Resettlement has developed a set of guidelines for an emergency cash transfer programme. However, the financial mechanism to fund such a programme is still lacking. Myanmar's new National Framework for Community Disaster Resilience suggests that the country's existing cash transfer programme, which provides payments to poor households with children, could be expanded to provide disaster-linked payments (Myanmar National Framework for Community Disaster Resilience, n.d.).

- *Philippines*: The Philippines does not have a disaster-linked cash transfer programme, but other programmes have been employed in recent disasters to identify some beneficiaries. In particular, the Philippines' large scale conditional cash transfer programme (Pantawid Pamilyang Pilipino Program or the 4Ps) – currently providing financial grants to 4.4 million low-income families with children – was used to quickly identify participants for cash-for-work programmes following Typhoon Haiyan in 2013 (World Bank, 2014b). To ensure comprehensive coverage of the agricultural sector, such a scheme would need to be expanded to include low-income households without children when disasters occur.

- *Thailand*: In addition to the agricultural insurance programme, Thailand also disburses small payments to all farmers in affected regions when an official disaster is declared. Farmers are eligible to receive THB 6 956 (USD 201) per ha of rice, THB 7 175 (USD 208) per ha of field crops and THB 10 563 (USD 306) per ha of horticulture crops (MoAC, 2017a). While potentially an important tool for helping farmers purchase new inputs, the scheme is slow to issue pay-outs (3-4 months on average) and thus decelerates the recovery process (Chantanusornsiri, 2016). Moreover, disasters declared at the aggregate level do not necessarily correspond with disasters suffered at the farm level. In other words, some farmers do not receive pay-outs when suffering from losses and others receive pay-outs that they did not need. Offering such payments to farms of all sizes also makes the scheme very expensive; larger farms could afford to purchase insurance or manage the financial risk of weather-related disasters through other means.

Table 4.3. Selected examples of disaster-linked cash transfer programme payments for agricultural producers in Viet Nam, 2017

	Per hectare		
	0-30% Damaged	30-70% Damaged	> 70% Damaged
Rice, maize and other cash crops	0	VND 1 mln (USD 44)	VND 2 mln (USD 88)
High-yield rice	0	VND 1.5 mln (USD 66)	VND 3 mln (USD 132)
Trees (e.g. coffee, rubber, fruit)	0	VND 2 mln (USD 88)	VND 4 mln (USD 176)
	Per head		
Poultry	VND 10-35 000 (USD 0.44-1.54)		
Buffalo, cow, horse	VND 0.5-6 mln (USD 22-264)		
Deer, sheep, goat	VND 1-2.5 mln (USD 44-110)		

Note: Under a previous decree (Decision 49/2012/QD-TTg), farmers often received less than the rates above, as local governments contributed to these payments and frequently lacked the necessary funds. The 2017 decree aims to address this issue by mobilising resources from a comprehensive range of sources and issuing a new supporting mechanism from the central budget (CIEM, 2017).
Source: Decree 02/2017/ND-CP dated 09 January 2017.

- *Viet Nam*: Viet Nam's disaster-linked cash transfer programme provides small payments to farmers according to the type of commodity and level of losses suffered. For instance, a farmer growing rice, maize and other cash crops qualifies for VND 2 million (USD 88) per hectare if damage is more than 70% and VND 1 million (USD 44) per hectare for damage between 30% and 70% (Table 4.3) (Viet Nam, 2012). When a farmer suffers from a loss due to a natural hazard, he or she can apply for an assessment of the farm's losses – a system that can be time-intensive and expensive. All farms are eligible to participate in this scheme regardless of size, further inflating the cost of the programme.

4.5. Recovery needs and implementation are fitted to each disaster circumstances but do not build resilience to future disasters

Recovery from a disaster involves policy measures that help populations building back better (OECD, 2014b). Recovery should not be limited to re-instating a pre-disaster situation although that is often the limited objective governments try to reach, even in OECD countries. Similarly, the ASEAN countries under study concentrate their disaster recovery efforts to bringing their economic potential back to where it was before the disaster hit. They increasingly use the Post-Disaster Needs Assesment (PDNA) method to estimate the needs to recover from weather-related disasters, which allows governments more efficiently to earmark the budget needed or to target technical and financial assistance from the international development community (UNESCAP, 2015; World Bank, 2011). This section reviews two key policy instruments to aid the agricultural sector in the recovery process: (i) provision of agricultural inputs and equipment and (ii) debt rescheduling measures.

Inputs or equipment can be slow to reach their beneficiaries if procurement processes are not streamlined

The policy guidelines from international organisations recommend setting up legal frameworks and institutions to implement effective DRM, which includes jump-starting the recovery process by providing timely inputs and equipment to resume agricultural production (ASEAN, 2016; Carter, 2008; CFE-DMHA, 2015; IFRC and UNDP, 2014; UN, 2015). An effective tool to accelerate the capacity of the agricultural sector to get back into production after a disaster is the agricultural component of the PDNA (ADB, 2013; FAO, 2016a, 2015b, 2015c, 2009)

In the four ASEAN countries studied, the ministry in charge of agriculture is usually responsible for sourcing and distributing essential commodities for the agricultural sector after it is hit by a disaster: rice, seeds, insecticides, small farming equipment and veterinary medicine. These items are sometimes held in the national reserve along with basic food, drinking water and other emergency commodities to be distributed to affected areas. The countries reviewed take this issue very seriously because leaving poor smallholder farmers to fend for themselves after a disaster would have strong negative impacts on the country's food security and social stability. In this sense, ASEAN countries are usually better equipped than OECD countries to supply inputs and equipment to their farmers. Nevertheless, the evidence gathered suggests that the efforts to provide inputs and equipment to farmers after a disaster could be streamlined and made more efficient with the help of local civil society and international development partners to reach even more farmers in areas affected by weather-related disasters.

The example of the Philippines and its PDNA after typhoons Ondoy and Pepeng shows how the rapid assessment of the needs to recover from a disaster can lead to streamlining of funds being identified and earmarked for recovery, which allows speedy rehabilitation and economic activity to resume promptly (World Bank, 2011). Similarly, Myanmar requested the technical assistance of FAO to prepare its agricultural PDNA after cyclone Nargis, which made it easier for the country and its development partners to source the funds needed for the recovery process (FAO, 2009).

Nevertheless, some government services still need to become more efficient in procuring the products needed for post-disaster recovery. Despite a USD 5 million loan from the World Bank to buy small agricultural equipment for farmers hit by catastrophic floods in 2015, the government of Myanmar was not able to spend the funds allocated quickly enough to foster a quick recovery (World Bank, 2017).

Debt rescheduling is a common financial assistance tool in many Southeast Asian countries

A useful financial tool to support the recovery process after an extreme weather event is debt rescheduling. Agricultural loans should include clear conditions under which debt is rescheduled. This provides farmers with a limited, implicit "insurance" against catastrophic risks and enables them to make productive investments. Without such schemes, the long-term impact of weather-related disasters can be severe. If farmers do not have sufficient financing to pay off their loans, they may be forced to sell their land (Kloeppinger-Todd and Sandar, 2013).

At the same time, such rescheduling terms should be conservative and write-offs for the interest or loan principals should be avoided. A key concern with large debt write-offs after disasters is that they may encourage farmers to make high-risk investments and increase vulnerability because they know that their debts will be written off in case of a disaster. Large write-offs may also crowd out private insurance providers from the insurance market by lowering the demand for insurance products and making it difficult for them to finance their own debt (OECD, 2011, 2009).

Debt rescheduling is a common financial assistance tool in Southeast Asia. Some countries employ such measures in a more systematic way than others. In some countries, interest can be written off for certain loan programmes, which may increase farmer vulnerability. In addition to these general regional findings, the following summarises the specific local situation and policy approaches to emergency cash transfers after a disaster in the four countries under study.

- *Myanmar*: While loan access in rural areas is relatively limited[10], the two main formal providers – the Myanmar Agriculture Development Bank (MADB) and MOALI's Co-operatives Department – offer debt rescheduling. The terms of MADB loans can be relaxed if farmers send an application after a disaster occurs and it is approved by an advisory committee. Interest can also be written off for a finite period. The Co-operatives Department also reschedules existing loan periods and issues additional six-month loans when a disaster occurs (MOALI, 2017).

- *Philippines*: Debt cannot be written off, but loan payments can be rescheduled and additional loans can be issued. "The Survival and Recovery Loan Program provides financing to supplement existing resources of the Department of Agriculture and the Agricultural Credit Policy Council in providing calamity loan assistance to farmer families in typhoon-affected areas for the rehabilitation of their livelihoods. The Agricultural Credit Policy Council provides zero-cost interest rate to eligible conduits under a fund management arrangement. In turn, these disburse loan funds to affected families at 0% interest. Eligible end-borrowers are small farmers or their household members who are existing borrowers" (OECD, 2017a).

- *Thailand*: The Thai government's commitments to reschedule debt or reduce interest rates are decided on an annual basis. For 2016-17, the government committed to reducing the interest of agricultural loans issued through two specific agencies, namely: (i) by covering 3% of the interest rate for Farmer Co-operative Members in Southern Thailand that were affected by drought and (ii) covering interest for one year for farmers of the Revolving Fund (MoAC, 2017a).

- *Viet Nam*: Debt rescheduling is common for agricultural loans (MARD, 2017b).

Notes

[1]   At the irrigation perimeter or community level, collective water management innovations have been trialled successfully in OECD countries: lining of irrigation canals, capturing and using rainwater, benchmarking among water suppliers to limit distributional channel losses, establishing an industry code of practice for irrigation system design and use. However, improving the efficiency of water use by farmers can have perverse effects and counterproductive outcomes by limiting the natural water recharge into groundwater systems, encouraging more irrigation, and fostering farms to specialise in the production of more water-thirsty crops (OECD, 2016c, 2015c).

[2]   In terms of monitoring of extension services, national statistics usually report the output of extension (e.g. number of training sessions given or number of farmers trained) but fail to account for extension outcomes like the level of awareness of farmers on weather-related risks and the adoption of farm practices to alleviate these risks.

[3]   Approximately 10% of farmers in Myanmar are co-operative members (MOALI, 2017).

[4]   For instance, drought insurance could include premium discounts to farmers that use water more efficiently. However, this would require individual water meters to measure water use by farmers or irrigation groups, which is not well developed in many countries.

[5]   An additional challenge with traditional crop insurance products is that their basic structure incentivises moral hazard. As pay-outs are determined by losses at the plot level, there is a risk that some farmers may make less effort or engage in maladaptive practices to intentionally disrupt yields and benefit from the insurance coverage.

[6]   For example, using the same model as the African Risk Capacity (www.africanriskcapacity.org/) and the Caribbean Catastrophe Risk Insurance Facility (www.ccrif.org/).

[7]   Flood or excessive rain, drought, frost, windstorm or typhoon, fire, hail, and damage by pests and diseases (Rattanayod, 2016).

[8]   Approximately 95% of Thai farmers are members

[9]   On strengthening agricultural statistics, the Asian Development Bank is piloting the use of satellites to collect land use and crop production estimates in the Lao People's Democratic Republic, the Philippines, Thailand and Viet Nam through a regional project on innovative data collection methods for agricultural and rural statistics (www.adb.org/projects/46399-

001/main#project-overview). This pilot project is expected to provide good statistical data gathering practices that could be emulated by other countries in the region.

[10] Myanmar's agricultural sector contributes 43% of GDP and employs 54% of the population; however, less than 3% of all outstanding loans are made to this sector. Estimates suggest 10% of rural areas in Myanmar have access to formal financial services. Barriers to expansion include regulatory constraints on commercial banks providing agricultural loans, geographic distance from farmers to bank branches in township centres and land ownership requirements (Kloeppinger-Todd and Sandar, 2013). According to non-governmental sources, private sector loans have declined due to a recent disaster; private loan providers incurred major losses when farmers defaulted on their loans and the government did not step in.

References

ADB (2013), *Myanmar: Agriculture, Natural Resources, and Environment Initial Sector Assessment, Strategy, and Road Map*, Asian Development Bank, Mandaluyong City, https://www.adb.org/documents/myanmar-agriculture-natural-resources-and-environment-initial-sector-assessment-strategy (accessed 28 February 2017).

ADRC Myanmar (2017), Information on Disaster Risk Reduction of the Member Countries Myanmar, http://www.adrc.asia/nationinformation.php?NationCode=104&Lang=en&Mode=country (accessed 29 March 2017).

ADRC Philippines (2017), Information on Disaster Risk Reduction of the Member Countries Philippines, http://www.adrc.asia/nationinformation.php?NationCode=608&Lang=en&NationNum=14 (accessed 29 March 2017).

ADRC Thailand (2017), Information on Disaster Risk Reduction of the Member Countries Thailand, http://www.adrc.asia/nationinformation.php?NationCode=764 (accessed 29 March 2017).

ADRC Viet Nam (2017), Information on Disaster Risk Reduction of the Member Countries Viet Nam, http://www.adrc.asia/nationinformation.php?NationCode=704&Lang=en&NationNum=15 (accessed 30 March 2017).

AFD (2014), "Lutter contre le changement climatique et étendre les surfaces irriguées le long du Fleuve rouge", Agence Française de Développement, Hanoi, http://www.afd.fr/home/pays/asie/geo-asie/afd-vietnam/cac-du-an-cua-afd-tai-viet-nam/developpment-rural-et-agricole/maitriser-les-eaux-du-fleuve-rouge (accessed 21 April 2017).

Alano, E. and M. Lee (2016), "Natural disaster shocks and macroeconomic growth in Asia: evidence for typhoons and droughts", *ADB Economics Working Paper Series*, No. 503, Asian Development Bank, Mandaluyong City, https://papers.ssrn.com/sol3/papers.cfm?abstract_id=2894778 (accessed 27 February 2017).

ASEAN (2016), *ASEAN Disaster Recovery Reference Guide*, ASEAN Secretariat, Jakarta.

Artemis (2017), "ASEAN countries explore regional risk insurance pool", Artemis website, http://www.artemis.bm/blog/2017/02/20/asean-countries-explore-regional-risk-insurance-pool/ (accessed 23 May 2017).

Australian Embassy to Viet Nam (2017), Interview with Australian Embassy to Viet Nam, Hanoi, 15 February 2017.

Aye, N.N. (2014), *Country report of Myanmar 2014*, Asian Disaster Reduction Center, Kobe, http://www.adrc.asia/countryreport/MMR/2014/FY2014A_MMR_CR.pdf (accessed 28 February 2017).

BAAC (2010), *BAAC Savings Mobilization*, Bank of Agriculture and Agricultural Cooperatives, Bangkok, http://www.apraca.org/upload/book/47/10-10%20okpaper%20savings%20mob%20baac.pdf (accessed 29 March 2017).

Bangko Sentral ng Pilipinas (2015), *National baseline survey on financial inclusion*, Central Bank of the Philippines, Manila, http://www.bsp.gov.ph/downloads/publications/2015/NBSFIFullReport.pdf (accessed 29 March 2017).

Beresnev N. and J. Broadhead (2016), *Financing for mangrove protection with emphasis on Pakistan, Thailand and Viet Nam*, FAO and IUCN, Bangkok, http://www.fao.org/3/a-i6524e.pdf (accessed 3 March 2017).

Carter, W.N. (2008), *Disaster management: a disaster manager's handbook*, Asian Development Bank, Mandaluyong City, https://www.adb.org/publications/disaster-management-disaster-managers-handbook (accessed 28 February 2017).

Carter, M.R., C.B. Barrett, S. Boucher, S. Chantarat, F. Galarza, J. McPeak, A.G. Mude and C. Trivelli (2008), "Insuring the never before insured: explaining index insurance through financial education games", *BASIS Brief* 2008-07, University of Wisconsin-Madison Department of Agricultural and Applied Economics, Madison, http://hdl.handle.net/10568/765 (accessed 22 May 2017).

Centre for Economic and Social Development (2017), Interview with Centre for Economic and Social Development, Yangon, 14 February 2017.

CFE-DMHA (2015), *ASEAN Disaster Management Reference Handbook*, Center for Excellence in Disaster Management & Humanitarian Assistance, Ford Island, Hawaii, http://reliefweb.int/sites/reliefweb.int/files/resources/ASEAN%20Disaster%20Management%20Reference%20Handbooks.pdf (accessed 1 March 2017).

Chaipattana Foundation, (2017), *Philosophy of Sufficiency Economy*, Chaipattana Foundation, Bangkok, Thailand, http://www.chaipat.or.th/chaipat_english/index.php?option=com_content&view=article&id=4103&Itemid=293 (accessed 3 April 2017).

Chantanusornsiri, W. (2016), "Thai governments are failing our farmers", *Bangkok Post* 30 June 2016, Bangkok, Thailand, http://www.bangkokpost.com/print/1023477/ (accessed 10 April 2017).

Chariyaphan, R. (2012), *Thailand's country profile 2012*, Asian Disaster Reduction Center, Kobe, Japan, http://www.adrc.asia/countryreport/THA/2012/THA_CR2012B.pdf (accessed 28 February 2017).

CIEM (2017), Interview with Central Institute for Economic Management, Hanoi, 16 February 2017.

Conway, G. (2012), *One Billion Hungry: Can We Feed the World?*, Cornell University Press, Ithaca, New York.

Cornell University (2017), SRI-Rice, SRI International Network and Resources Center, College of Agriculture and Life Sciences, Cornell University, Ithaca, New York, http://sri.ciifad.cornell.edu/ (accessed 12 April 2017).

Economics of Land Degradation Initiative (2016) "Sustainable Land Management: Adoption and Implementation Constraints", *EEPSEA Project Brief*, The Economy and Environment Program for Southeast Asia, Los Baños, Philippines, http://www.eepseapartners.org/wp-content/uploads/pdfs/2016/ELD-primer7.pdf (accessed 29 March 2017).

EEPSEA (2017) Interview with Economy and Environment Program for Southeast Asia (EEPSEA), Bangkok, 7 February 2017.

European Commission (2011), *Towards Better Environmental Options for Flood Risks Management*, European commission, Brussels, Belgium, http://ec.europa.eu/environment/water/flood_risk/better_options.htm (accessed 24 April 2017).

FAO (2016a), *"El Niño" event in Viet Nam: agriculture, food security and livelihood needs assessment in response to drought and salt water intrusion*, FAO, Hanoi, Viet Nam, http://www.fao.org/3/a-i6020e.pdf (accessed 3 March 2017).

FAO (2016b), Proceedings of the Workshop on Forests and Disasters in Southeast Asia, 24-25 August 2016, Antipolo City, Philippines, FAO, Bangkok, Thailand.

FAO (2015a), *The Impact of Disasters on Agriculture and Food Security*, FAO, Rome, Italy, http://www.fao.org/resilience/resources/resources-detail/en/c/346258/ (accessed 29 March 2017).

FAO (2015b), *2015–2016 El Niño - Early action and response for agriculture, food security and nutrition - Update #4*, FAO, Rome, Italy, http://www.fao.org/resilience/resources/resources-detail/en/c/340660/ (accessed 29 March 2017).

FAO (2009), *Post-Nargis Recovery and Rehabilitation Programme*, FAO, Yangon, Myanmar, http://www.fao.org/fileadmin/user_upload/emergencies/docs/FAO_Myanmar__Post_Nargis_Recovery_and_Rehabilitation_Programme_Strategy.pdf (accessed 29 March 2017).

FAO-MM (2017), Interview with FAO Representation to Myanmar, Yangon, 10 February 2017.

FAORAP (2017), Interview with FAO Regional Office for Asia and the Pacific, Bangkok, 9 February 2017.

FAO-VN (2017), Interview with FAO Representation to Viet Nam, Hanoi, 16 February 2017.

FCRI (2017), Interview with Field Crops Research Institute, Hanoi, 16 February 2017.

GFDRR (2012), *Advancing Disaster Risk Financing and Insurance in ASEAN Member States: Framework and Options for Implementation*, Global Facility for Disaster Reduction and Recovery, World Bank, Washington, DC, https://www.gfdrr.org/sites/gfdrr/files/publication/DRFI_ASEAN_REPORT_June12.pdf (accessed 23 May 2017).

GiZ ASEAN SAS (2017), Interview with ASEAN SAS, GiZ, Yangon, 10 February 2017.Htoo Thant (2016), "Irrigation budget halved in agricultural policy change", *Myanmar Times*, 20 July 2016, http://www.mmtimes.com/index.php/national-news/21475-irrigation-budget-halved-in-agricultural-policy-change.html (accessed 24 March 2017).

IFRC and UNDP (2014), *Effective law and regulation for disaster risk reduction: a multi-country report*, IFRC and UNDP, New York, New York, http://www.undp.org/content/undp/en/home/librarypage/crisis-prevention-and-recovery/effective-law---regulation-for-disaster-risk-reduction.html (accessed 28 February 2017).

IPSARD (2017), Interview with IPSARD, Hanoi, 16 February 2017.

IPSARD (2011), "Vietnam Food Security and Rice Value Chain. Vietnam's Rice Balance: Recent Trends, Future Projections, and Implications for Policy", *Collaborative Research Program Policy Note* No. 1, IPSARD, Hanoi, Viet Nam.

IRRI (2017), Saving water with Alternate Wetting Drying (AWD), International Rice Research Institute, Los Baños, Philippines, http://www.knowledgebank.irri.org/index.php?option=com_zoo&view=item&layout=item&Itemid=474 (accessed 12 April 2017).

IWMI (2015), Improving water management in Myanmar's Dry Zone for food security, livelihoods and health, International Water Management Institute, Yangon, Myanmar, http://www.iwmi.cgiar.org/Publications/Other/Reports/PDF/improving-water-management-in-myanmars-dry-zone-for-food-security-livelihoods-and-health.pdf (accessed 5 April 2017).

Kloeppinger-Todd, R. and T.M. Sandar (2013), "Rural Finance in Myanmar", Background Paper No.3, Strategic Agricultural Sector and Food Security Diagnostic for Myanmar, Michigan State University and Myanmar Development Resource Institute - Centre for Economic and Social Development,

Yangon, Myanmar,
http://fsg.afre.msu.edu/Myanmar/myanmar_background_paper_3_rural_finance.pdf (accessed 29 March 2017).

Land Development Department (2017), *Soil Doctors Volunteer*, Thai Land Development Department, Bangkok, Thailand, http://www.ldd.go.th/ldd_en/en-US/soil-doctor-volunteer/ (accessed 29 March 2017).

Losch, B. (Ed) (1996), *Les agricultures des zones tropicales humides : éléments de réflexion pour l'action*, Ministère de la coopération, Paris, France.

Lotsch, A., W. Dick, O.P. Manuamorn (2010), "Assessment of innovative approaches for flood risk management and financing in agriculture", *Agriculture and Rural Development Discussion Paper* 46, World Bank, Washington, DC, http://siteresources.worldbank.org/INTARD/Resources/Assessment_Combind_Web_small.pdf (accessed 24 May 2017).

MARD (2017a), *Report: Updates policies to support agriculture in 2016 – 2017*, Ministry of Agriculture and Rural Development, Hanoi, Viet Nam.

MARD (2017b), Interview with Ministry of Agriculture and Rural Development, Hanoi, 15 February 2017.

MoAC (2017a), *Disaster Preparedness Plan to Mitigate Drought Effect 2016-2017*, Ministry of Agriculture and Cooperatives, Bangkok, Thailand.

MoAC (2017b), Interview with Ministry of Agriculture and Cooperatives, Bangkok, 9 February 2017.

MOALI (2017), Interview with Ministry of Agriculture, Livestock and Irrigation, Nay Pyi Taw, 13 February 2017.

MONRE (2014), "Establishing Disaster Prevention Fund: Active response", Press release 28 November 2014, Ministry of Natural Resources and Environment, Hanoi, Viet Nam, http://reliefweb.int/report/viet-nam/establishing-disaster-prevention-fund-active-response (accessed 30 March 2017).

Myanmar (2016), Lessons Learnt Workshop on Myanmar Action Plan on Disaster Risk Reduction, (MAPDRR) Implementation, 8th April 2016, Nay Pyi Taw, Myanmar.

Myanmar (2012), *Myanmar Action Plan on Disaster Risk Reduction (MAPDRR)*, The Republic of the Union of Myanmar, Nay Pyi Taw, Myanmar.

Myanmar National Framework for Community Disaster Resilience (n.d.), Draft Myanmar National Framework for Community Disaster Resilience, received February 2017.

Nabangchang, O., H. Francisco and A. Zakaria (2014), "Thailand's 2011 Flood: Household Damages, Compensation and Natural Catastrophe Insurance", *China-USA Business Review*.

Nguyen, K.A. (2017), "Aquaculture insurance pilot program in Viet Nam", a presentation given to the *GiZ workshop Study on the lessons learned from the National Agriculture Insurance Pilot Programme in Viet Nam*, 2 March 2017, Hanoi, Viet Nam.

OECD (2017a), *Agricultural Policies in the Philippines*, OECD Publishing, Paris, http://dx.doi.org/10.1787/9789264269088-en.

OECD (2017b), *Agricultural Policy Monitoring and Evaluation 2017*, OECD Publishing, Paris, http://dx.doi.org/10.1787/agr_pol-2017-en.

OECD (2017c), *Building Food Security and Managing Risk in Southeast Asia*, OECD Publishing, Paris, http://dx.doi.org/10.1787/9789264272392-en

OECD (2017d), *OECD Recommendation on Disaster Risk Financing Strategies*, OECD Publishing, Paris, http://www.oecd.org/finance/oecd-recommendation-disaster-risk-financing-strategies.htm

OECD (2016a), *Agricultural Policy Monitoring and Evaluation 2016*, OECD Publishing, Paris. DOI: http://dx.doi.org/10.1787/agr_pol-2016-en

OECD (2016b), *Agriculture and Water* OECD Meeting of Agriculture Ministers Background Note http://www.oecd.org/tad/sustainable-agriculture/5_background_note.pdf

OECD (2016c), *Mitigating Droughts and Floods in Agriculture: Policy Lessons and Approaches*, OECD Publishing, Paris, http://dx.doi.org/10.1787/9789264246744-en.

OECD (2016d), *Trends in Risk Communication Policies and Practices*, OECD Publishing, Paris, http://dx.doi.org/10.1787/9789264260467-en

OECD (2015a), *Agricultural Policies in Viet Nam 2015*, OECD Publishing, Paris, http://dx.doi.org/10.1787/9789264235151-en.

OECD (2015b), "A Health Check for Water Resources Allocation", in *Water Resources Allocation: Sharing Risks and Opportunities*, OECD Publishing, Paris, http://dx.doi.org/10.1787/9789264229631-9-en.

OECD (2015c), *Drying Wells, Rising Stakes: Towards Sustainable Agricultural Groundwater Use*, OECD Publishing, Paris, http://dx.doi.org/10.1787/9789264238701-en.

OECD (2015d), *The Changing Face of Strategic Crisis Management*, OECD Publishing, Paris, http://dx.doi.org/10.1787/9789264249127-en.

OECD (2014a), *Boosting Resilience through Innovative Risk Governance*, OECD Publishing, Paris, http://dx.doi.org/10.1787/9789264209114-en

OECD (2014b), *Recommendations on the Governance of Critical Risks,* OECD Publishing, Paris, http://www.oecd.org/gov/risk/recommendation-on-governance-of-critical-risks.htm

OECD (2011), *Managing Risk in Agriculture: Policy Assessment and Design*, OECD Publishing, Paris, http://dx.doi.org/10.1787/9789264116146-en.

OECD (2010), *Sustainable Management of Water Resources in Agriculture*, OECD Publishing, Paris, http://dx.doi.org/10.1787/9789264083578-en.

OECD (2009), *Managing Risk in Agriculture: A Holistic Approach*, OECD Publishing, Paris, http://dx.doi.org/10.1787/9789264075313-en.

Ortega, A.P. (2014), *Philippines' country report 2014*, Asian Disaster Reduction Center, Kobe, Japan, http://www.adrc.asia/countryreport/PHL/2014/PHL_CR2014B.pdf (accessed 28 February 2017).

Oxford Business Group (2016), *Agricultural insurance in Thailand shifts focus to crop protection*, Oxford Business Group, Oxford, UK, https://www.oxfordbusinessgroup.com/analysis/agricultural-insurance-crop-protection-top-priority-difficult-weather-conditions-promise-impact (accessed 29 March 2017).

Patmasiriwat, D. and S. Hengpatana (2016), "Income, saving, and wealth of Thai rural households: a case study of saving adequacies", *Applied Economics Journal*, Vol. 23, No.1, pp. 75-91.

Peel, M. (2016), "Thailand offers $1bn loan to rice farmers in push to boost prices", *Financial Times*, 31 October 2016, Financial Times, London, United Kingdom.

Quang, N.M. (2017), "Is Vietnam in for Another Devastating Drought?", *The Diplomat*, 8 February 2017, http://thediplomat.com/2017/02/is-vietnam-in-for-another-devastating-drought/ (accessed 2 March 2017).

Rattanayod, B. (2016), "Sovereign Risk Transfer Solutions Thailand", Presentation to the GiZ Executive Consultation and Capacity Building Seminar for Asian Government Representatives, 17-19 October 2016, Jakarta, Indonesia, GiZ, Bangkok, Thailand, https://jakarta-2016.climate-risk-insurance.org/ (accessed 5 April 2017)

Richards, D.R. and D.A. Friess (2016), "Rates and drivers of mangrove deforestation in Southeast Asia, 2000-2012", *Proceedings of the National Academy of Sciences* USA, Vol. 113, pp. 344-349.

RID (2017), Interview with Royal Irrigation Department, Bangkok, 9 February 2017.

SEARCA (2017), Interview with SEARCA, Paris, 27 January 2017.

Seng, L.K. (2016), "Emergency situations, participation, and community-based disaster responses in Southeast Asia: gray areas and causes for optimism", *Philippine Studies: Historical and Ethnographic Viewpoints*, Vol. 64, No. 3-4, pp. 499-526.

Surminski, S. (2013), "The Role of Insurance in Reducing Direct Risk — The Case of Flood Insurance" *International Review of Environmental and Resource Economics*, Vol. 7, pp. 241–278.

Swiss Capacity Building Facility (2015), *Feasibility Study and Dry Run for Agricultural Insurance in Myanmar*, Swiss Capacity Building Facility, Zurich, Switzerland, http://scbf.ch/wp-content/uploads/2015/10/SCBF_FSW-15_FactSheet_SFSA-Myanmar.pdf (accessed 29 March 2017).

Tinh, T.V. (2017), "Study on the lessons learned from the national agriculture insurance pilot program in Vietnam: International experiences and development of agricultural insurance in Vietnam; Evaluation of rice and livestock insurance", presentations given to the *GiZ workshop Study on the lessons learned from the National Agriculture Insurance Pilot Programme in Viet Nam*, 2 March 2017, Hanoi, Viet Nam.

Tong Y.D. (2015), "A cost-benefit analysis of dike heightening in the Mekong Delta", *EPSEA Report* 2015-RR11, Economy and Environment Program for Southeast Asia, Los Baños, Philippines, http://www.eepsea.org/pub/rr/2015-RR11_Tong_digital_final.pdf (accessed 31 March 2017).

Trethewie, S. (2013), "The ASEAN Plus Three Emergency Rice Reserve (APTERR): Cooperation, commitment and contradictions", *NTS Working Paper* no. 8, RSIS Centre for Non-Traditional Security (NTS) Studies, Singapore, https://www.rsis.edu.sg/wp-content/uploads/rsis-pubs/NTS/resources/research_papers/NTS_Working_Paper8.pdf (accessed 12 April 2017).

UN (2015), *Sendai Framework for Disaster Risk Reduction 2015 – 2030*, United Nations, New York, New York, http://www.preventionweb.net/files/43291_sendaiframeworkfordrren.pdf (accessed 29 March 2017).

UNDP (2017a), Interview with UNDP Bangkok Regional Hub, Bangkok, 6 February 2017.

UNDP (2017b), "Policy brief for biodiversity finance initiative in Thailand", BIOFIN Thailand, UNDP, Bangkok, Thailand.

UNDP-TH (2017), Interview with UNDP Thailand, Bangkok, 6 February 2017.

UNESCAP (2015) "Financing Disaster Risk Reduction for sustainable development in Asia and the Pacific", Draft Discussion Paper for the Asia-Pacific High-Level Consultation on Financing for Development, Jakarta, Indonesia 29-30 April 2015, UNESCAP, Bangkok, Thailand, https://www.unescap.org/sites/default/files/8-ESCAP_FfD_Financing%20Disaster_24April2015.pdf (accessed 29 March 2017).

UNISDR (2013a), *Country Assessment Report for the Philippines. Strengthening of Hydrometeorological Services in Southeast Asia*, UNISDR, Geneva, Switzerland, http://www.unisdr.org/we/inform/publications/33988 (accessed 3 March 2017).

UNISDR (2013b), *Country Assessment Report for Viet Nam. Strengthening of Hydrometeorological Services in Southeast Asia*, UNISDR, Geneva, Switzerland, http://www.unisdr.org/we/inform/publications/33988 (accessed 3 March 2017).

UNISDR (2008), *Indigenous Knowledge for Disaster Risk Reduction: Good Practices and Lessons Learned from Experiences in the Asia-Pacific Region*, UNISDR, Bangkok, Thailand, https://www.unisdr.org/we/inform/publications/3646 (accessed 28 February 2017).

USDA (2017), *Rice Market and Policy Changes over the Past Decade*, USDA Foreign Agricultural Service, Bangkok, Thailand, https://www.fas.usda.gov/data/thailand-rice-market-and-policy-changes-over-past-decade (accessed 2 March 2017).

US Embassy to Viet Nam (2017), Interview with USAID and USDA Representation to Viet Nam, Hanoi, 17 February 2017.

Viet Nam (2012), Prime Minister Decision No 49/1012, 08 November 2012, Hanoi, Viet Nam.

Ward, P.S., D.J. Spielman, D.L. Ortega, N. Kumar and S. Minocha (2015), "Demand for complementary financial and technological tools for managing drought risk", *IFPRI Discussion Paper* 1430, International Food Policy Research Institute, Washington, DC, http://ebrary.ifpri.org/cdm/ref/collection/p15738coll2/id/129092 (accessed 24 May 2017).

World Bank (2017), Interview with World Bank, Yangon, 10 February 2017.

World Bank (2014a), *Financial Protection against Natural Disasters: An Operational Framework for Disaster Risk Financing and Insurance*, World Bank, Washington, DC, https://openknowledge.worldbank.org/handle/10986/21725 (accessed 29 March 2017).

World Bank (2014b), "Philippines: Social Protection Systems Help Mitigate Disaster and Climate Risk", *Press release 4 November 2014*, World Bank, Washington, DC, http://www.worldbank.org/en/news/press-release/2014/11/04/philippines-social-protection-systems-help-mitigate-disaster-and-climate-risk (accessed 29 March 2017).

World Bank (2013), "Building Resilience to Disaster and Climate Change through Social Protection", *Working Paper 79621*, World Bank, Washington, DC, http://documents.worldbank.org/curated/en/187211468349778714/Building-resilience-to-disaster-and-climate-change-through-social-protection (accessed 29 March 2017).

World Bank (2012a), "Weather Based Crop Insurance in India", *Policy Research Working Paper 5985*, World Bank, Washington, DC, http://elibrary.worldbank.org/doi/pdf/10.1596/1813-9450-5985 (accessed 29 March 2017).

World Bank (2012b), The World Bank and Irrigation, World Bank, Washington, DC., http://lnweb90.worldbank.org/oed/oeddoclib.nsf/DocUNIDViewForJavaSearch/D09BF623218371348525680E00642EA2 (accessed 12 April 2017).

World Bank (2011), *Typhoon Ondoy and Pepeng post disaster needs assessment*, World Bank, Washington, DC, http://documents.worldbank.org/curated/en/551271468296967206/Executive-summary (accessed 29 March 2017).

ORGANISATION FOR ECONOMIC CO-OPERATION AND DEVELOPMENT

The OECD is a unique forum where governments work together to address the economic, social and environmental challenges of globalisation. The OECD is also at the forefront of efforts to understand and to help governments respond to new developments and concerns, such as corporate governance, the information economy and the challenges of an ageing population. The Organisation provides a setting where governments can compare policy experiences, seek answers to common problems, identify good practice and work to co-ordinate domestic and international policies.

The OECD member countries are: Australia, Austria, Belgium, Canada, Chile, the Czech Republic, Denmark, Estonia, Finland, France, Germany, Greece, Hungary, Iceland, Ireland, Israel, Italy, Japan, Korea, Latvia, Luxembourg, Mexico, the Netherlands, New Zealand, Norway, Poland, Portugal, the Slovak Republic, Slovenia, Spain, Sweden, Switzerland, Turkey, the United Kingdom and the United States. The European Union takes part in the work of the OECD.

OECD Publishing disseminates widely the results of the Organisation's statistics gathering and research on economic, social and environmental issues, as well as the conventions, guidelines and standards agreed by its members.

OECD PUBLISHING, 2, rue André-Pascal, 75775 PARIS CEDEX 16
(22 2018 01 1 P) ISBN 978-92-64-12352-6 – 2018